Daily *warm-ups*

PRE-ALGEBRA

by Hope Martin

J. WESTON
WALCH
PUBLISHER
Portland, Maine

Purchasers of this book are granted the right to reproduce all pages.
This permission is limited to a single teacher, for classroom use only.

Any questions regarding this policy or requests to purchase further
reproduction rights should be addressed to:

Permissions Editor
J. Weston Walch, Publisher
321 Valley Street • P.O. Box 658
Portland, Maine 04104-0658

1   2   3   4   5   6   7   8   9   10
ISBN 0-8251-4497-3
Copyright © 2003
J. Weston Walch, Publisher
P.O. Box 658 • Portland, Maine 04104-0658
www.walch.com
Printed in the United States of America

Table of Contents

The *Daily Warm-Ups* series is a wonderful way to turn extra classroom minutes into valuable learning time. The 180 quick activities—one for each day of the school year—review, practice, and teach pre-algebra. These daily activities may be used at the very beginning of class to get students into learning mode, near the end of class to make good educational use of that transitional time, in the middle of class to shift gears between lessons—or whenever else you have minutes that now go unused. In addition to providing students with structure and focus, they are a natural path to other classroom activities involving math or critical thinking.

As they solve these daily puzzles, your students will be using context clues and will interpret quantitative clues by making connections between abstract rules of number theory and motivating historical events. Each of the puzzles has one extra clue. This clue serves two purposes; the first is to give the students a substitute clue (in case they don't understand all of the puzzle-clues). Second, this extra clue encourages students to check their work. It helps them determine if the answer they found makes sense.

Daily Warm-Ups are easy-to-use reproducibles—simply photocopy the day's activity and distribute it. Or make a transparency of the activity and project it on the board. You may want to use the activities for extra credit points or as a check on the math skills that are built and acquired over time.

However you choose to use them, *Daily Warm-Ups* are a convenient and useful supplement to your regular lesson plans. Make every minute of your class time count!

States of the Union

The growth of the United States took place over the course of 172 years. Delaware was the first of the colonies to join the Union (December 7, 1787) and Hawaii became the 50th state (August 21, 1959).

The mathematics puzzles in this chapter help celebrate the nation we are today. When students solve the mathematical clues, they learn the year that each state became a part of the United States of America. Interdisciplinary discussions might encourage students to explore the correlation between the dates colonies and territories became states, life in colonial times, pioneer exploration, and the growth of the western part of the United States.

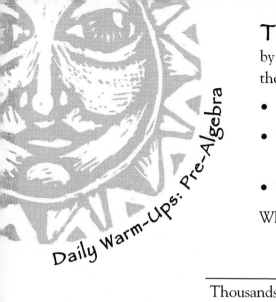

The Constitution of the United States was adopted by a Convention of the States on September 17 of this year. Solve the puzzle to find the year.

- My hundreds and units digits are the s~~ixth~~ ^{3rd} prime number.

- The two-digit number formed by my thousands and tens digits is equal to $3^2 \times 2$.

- The sum of all my digits is equal to $5^2 - 2$.

What year am I?

_____   _____   _____   _____
Thousands   Hundreds   Tens   Units

Delaware, Pennsylvania, and New Jersey were
the first three colonies to enter the Union. They all became states
in the month of December of the same year. Solve this puzzle to
learn the year.

- My hundreds and units digits could be the sides of a square with
  a perimeter of 28 units and an area of 49 square units.

- My tens digit is equal to 2^3.

- The sum of all of my digits is 2 less than the number of years
  in $\frac{1}{4}$ of a century.

What year am I?

Thousands	Hundreds	Tens	Units
_____	_____	_____	_____

Georgia, Massachusetts, Connecticut, New Hampshire, South Carolina, Virginia, New York, and Maryland all became states in the same year. They became the 4th through the 11th states of our United States. Solve this puzzle to learn the year.

- My tens and units digits form a number that is divisible by 2, 4, 8, 11, 22, and 44.

- My hundreds digit is the largest single-digit prime number.

- The sum of my digits is the same as the number of quarts in 6 gallons.

What year am I?

| _____ | _____ | _____ | _____ |
| Thousands | Hundreds | Tens | Units |

North Carolina became the 12th colony to join the Union. It became a state on November 21 of this year. To learn the year, just solve this puzzle.

- My hundreds, tens, and units digits are consecutive integers with a mean of 8.

- The sum of all of my digits is equal to the sum of the first five odd numbers.

What year am I?

_____ _____ _____ _____

Thousands   Hundreds   Tens   Units

4

Rhode Island was the last of the 13 original colonies to join the Union. It became a state on May 29. To find the year that Rhode Island became a state, just solve this puzzle.

- My date is divisible by 2, 5, and 10.

- My hundreds and tens digits are consecutive odd integers with a sum of 16.

- My hundreds digit is equal to $2^3 - 1$.

- The sum of all of my digits is the same as the two-digit number formed by my thousands and hundreds digits.

What year am I?

_____   _____   _____   _____
Thousands   Hundreds   Tens   Units

On March 4 of this year, Vermont became the 14th state to join the Union. To learn the year that Vermont became a state, just solve this puzzle.

- My hundreds and tens digits are odd numbers whose sum is 16 and product is 63.

- My hundreds digit is a prime number.

- The sum of all of my digits is the same as the number of months in $1\frac{1}{2}$ years.

What year am I?

| _____ | _____ | _____ | _____ |
| Thousands | Hundreds | Tens | Units |

Kentucky became the 15th state of the Union on June 1 of this year. To learn the year Kentucky gained statehood, just solve this puzzle.

- My units digit is the only even prime number.

- My hundreds and tens digits could be the sides of a rectangle with a perimeter of 32 units and an area of 63 square units.

- The sum of all of my digits is the largest prime number less than 20.

What year am I?

_____ _____ _____ _____
Thousands   Hundreds   Tens   Units

Tennessee became the 16th state of the United States when it joined the Union on June 1 of this year. To learn the year Tennessee became a state, just solve this puzzle.

- My units digit is a perfect number.

- The sum of my thousands and tens digit is the same as the number of years in a decade.

- My hundreds digit is 3 more than $\sqrt{16}$.

- The sum of all of my digits is the same as the first prime number greater than 20.

What year am I?

_____ _____ _____ _____
Thousands   Hundreds   Tens   Units

On March 1 of this year, Ohio became the 17th state of the United States. To learn the year, just solve this puzzle.

- My units digit is an odd number, it's prime, and it is the fourth number in the Fibonacci sequence.

- The two-digit number formed by my hundreds and tens digit is equal to 3 + 5 + 7 + 9 + 11 + 13 + 15 + 17.

- The sum of all of my digits is equal to $\sqrt{144}$.

What year am I?

_____ _____ _____ _____
Thousands   Hundreds   Tens   Units

Louisiana became the 18th state of the Union on April 30 of this year. To learn the year that Louisiana became a state, just solve this puzzle.

- The two-digit number formed by tens and units digits could be the perimeter of this rectangle:

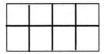

- My hundreds digit is 4 times my units digit.

- The three-digit number formed by my thousands, hundreds, and tens digits is a palindrome.

What year am I?

_____   _____   _____   _____
Thousands   Hundreds   Tens   Units

© 2003 J. Weston Walch, Publisher

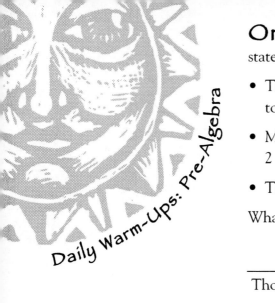

On December 11 of this year, Indiana became the 19th state to join the Union. Solve this puzzle to learn the year.

- The two-digit number formed by my tens and units digit is equal to 2^4.

- My hundreds digit is the same as the number of quarts in 2 gallons.

- The sum of all of my digits is equal to 2^4.

What year am I?

_____   _____   _____   _____

Thousands   Hundreds   Tens   Units

Mississippi became a state on December 10 of this year. It was the 20th state to join the Union. Solve this puzzle to learn the year.

- The two-digit number formed by my hundreds and units digits is 13 less than 10^2.

- The sum of all of my digits is 2 greater than the product of 5 and 3.

What year am I?

_____ _____ _____ _____
Thousands   Hundreds   Tens   Units

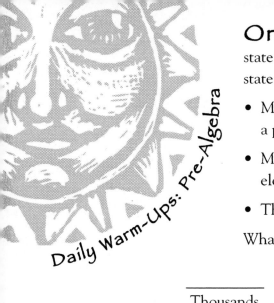

Daily Warm-Ups: Pre-Algebra

On December 3 of this year, Illinois became the 21st state to join the Union. To learn the year that Illinois became a state, just solve this puzzle.

- My hundreds and units digits could be the sides of a square with a perimeter of 32 units.

- My thousands and tens digits are the multiplicative identity element.

- The sum of all of my digits is divisible by 2, 3, 6, and 9.

What year am I?

_____	_____	_____	_____
Thousands	Hundreds	Tens	Units

Alabama became the 22nd state of the Union on December 14 of this year. To learn the year that Alabama became a state, just solve this puzzle.

- The two-digit number formed by my tens and units digits is the largest prime number less than 20.

- The two-digit number formed by my thousands and hundreds digits could be the area of this triangle:

- The sum of all of my digits is 19.

What year am I?

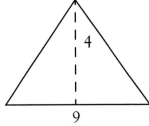

4

9

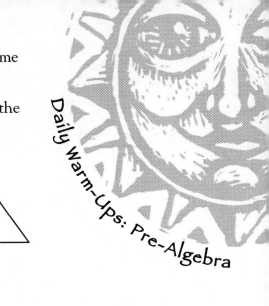

14

_____ _____ _____ _____
Thousands   Hundreds   Tens   Units

On March 15 of this year, Maine became the 23rd state of the United States. To learn the year this happened, just solve this puzzle.

- The two-digit number formed by my tens and units digits is the same as the number of faces on an icosahedron.

- My hundreds digit is 400% of my tens digit.

- The sum of all of my digits is a number that looks the same when it is written upside down.

What year am I?

——————— ——————— ——————— ———————
Thousands   Hundreds   Tens   Units

15

Missouri became the 24th state on August 10 of this year. To learn when Missouri became a state, just solve this puzzle.

- My hundreds digit is 400% of my tens digit.

- My tens digit is equal to $\sqrt[3]{8}$.

- If my tens digit were 6 greater, my date would be a palindrome.

What year am I?

| _____ | _____ | _____ | _____ |
| Thousands | Hundreds | Tens | Units |

16

On June 10 of this year, Arkansas became the 25th state of the United States. Solve this puzzle to learn the year.

- My tens digit is $\frac{1}{2}$ of my units digit; my hundreds digit is $1\frac{1}{3}$ times larger than my units digit.

- My units digit is a perfect number.

- The prime factors of the sum of all of my digits is 2×3^2.

What year am I?

_____ _____ _____ _____
Thousands   Hundreds   Tens   Units

17

Michigan became the 26th state of the United States on January 26 of this year. Solve this puzzle to learn the year.

- The two-digit number formed by my tens and units digits is the largest prime number between 30 and 40.

- The sum of my thousands and hundreds digits is 1 less than the sum of my tens and units digits.

- The sum of all of my digits is the largest prime number less than 20.

What year am I?

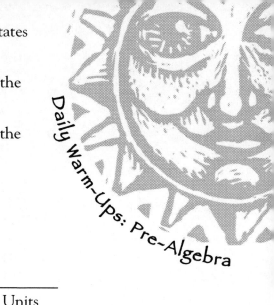

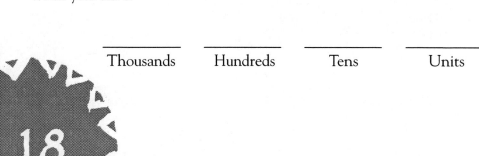

_____   _____   _____   _____

Thousands   Hundreds   Tens   Units

18

In this year, Florida and Texas became the 27th and 28th states of the United States. Solve this puzzle to learn the year.

- My hundreds digit is 200% of my tens digit.

- The two-digit number formed by my tens and units digits could be the area of this trapezoid:

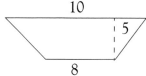

- The sum of all of my digits is equal to $\dfrac{2^3 + 7 \times 4}{2}$.

What year am I?

_____ _____ _____ _____

Thousands   Hundreds   Tens   Units

© 2003 J. Weston Walch, Publisher

On December 28 of this year, Iowa became the 29th state of the United States. Solve this puzzle to learn the year.

- My tens digit is $\frac{1}{2}$ of my hundreds digit and $\frac{2}{3}$ of my units digits.

- My tens digit is a square number; my hundreds digit is a cubic number.

- The sum of all of my digits is the largest prime number less than 20.

What year am I?

_____   _____   _____   _____
Thousands   Hundreds   Tens   Units

Wisconsin became the 30th state of the United States on May 29 of this year. To learn the year Wisconsin became a state, just solve this puzzle.

- The three-digit number formed by my hundreds, tens, and units digits is a palindrome.

- My hundreds digit is equal to 2^3; my tens digit is 50% of that.

- The sum of all of my digits is the number of days in three weeks.

What year am I?

_____	_____	_____	_____
Thousands	Hundreds	Tens	Units

On September 9 of this year, California became the 31st state of the United States. Solve this puzzle to learn the year.

- My date is a multiple of 10.
- My hundreds digit is a cubic number.
- My tens digit is 3 less than my hundreds digit.
- The sum of all of my digits could be the perimeter of a square with sides of length 3.5.

What year am I?

_____ _____ _____ _____
Thousands   Hundreds   Tens   Units

On May 11 of this year, Minnesota became the 32nd state of the United States. Solve this puzzle to learn the year.

- My hundreds and units digits could be the sides of a square with a perimeter of 32 units and an area of 64 square units.

- My tens digit is a prime number.

- The sum of all of my digits is a multiple of 11.

What year am I?

_____   _____   _____   _____
Thousands   Hundreds   Tens   Units

Oregon became the 33rd state of the United States on February 14 of this year. Solve this puzzle to learn the year.

- The two-digit number formed by my tens and units digits is the largest prime number between 50 and 60.

- The sum of my thousands and hundreds digits is equal to my units digit.

- Using order of operations, the sum of my digits is equal to $5 + 4 \times 2 + 10$.

What year am I?

_____   _____   _____   _____

Thousands   Hundreds   Tens   Units

24

Kansas became the 34th state of the United States on January 29 of this year. To learn the year this event took place, just solve this puzzle.

- If my tens digit were 2 greater, my date would be a palindrome.
- The sum of all of my digits is equal to 2^4.

What year am I?

_____ _____ _____ _____
Thousands   Hundreds   Tens   Units

West Virginia became the 35th state of the United States on June 20 of this year. To learn the year of this event, just solve this puzzle.

- My prime units digit is 50% of my tens digit.

- My tens digit is equal to 3!.

- The sum of all of my digits is equal to $\dfrac{16 + 5(2) + \sqrt{100}}{2}$.

What year am I?

_____   _____   _____   _____

Thousands   Hundreds   Tens   Units

26

On October 31 of this year, Nevada became the 36th state of the United States. To learn the year this happened, just solve this puzzle.

- The two-digit number formed by my thousands and hundreds digits is equal to my tens digit times 3.

- My hundreds, tens, and units digits are consecutive even integers, in reverse order.

- The sum of all of my digits is the same as the number of ounces in 1 pound, 3 ounces.

What year am I?

_____   _____   _____   _____
Thousands   Hundreds   Tens   Units

27

Nebraska, our 37th state, joined the Union on March 1 of this year. To learn the year Nebraska became a state, just solve this problem.

- The two-digit number formed by my tens and units digits is the largest prime number between 60 and 70.

- The three-digit number formed by my thousands, hundreds, and tens digits is equal to $10^2 + 9^2 + \sqrt{25}$.

- The sum of all of my digits is equal to the product of the first and fifth prime numbers.

What year am I?

_____   _____   _____   _____
Thousands   Hundreds   Tens   Units

Colorado became the 38th state of the Union on August 1 of this year. To learn the year this happened, just solve this puzzle.

- My hundreds, tens, and units digits are consecutive integers with a mean of 7; they are in reverse order.

- The sum of all of my digits is equal to the number of pints in 11 quarts.

What year am I?

_____ _____ _____ _____
Thousands   Hundreds   Tens   Units

29

In this year, four more states were added to the Union— North Dakota, South Dakota, Montana, and Washington. North and South Dakota were admitted on the same day, November 2; Montana followed on November 8; and Washington was admitted three days later. Solve this puzzle to learn the year.

- My hundreds and tens digits are the same; they are the sixth number in the Fibonacci sequence.

- My units digit is equal to the sum of the first three odd numbers.

- The sum of all of my digits is the same as the number of letters in the English alphabet.

What year am I?

_____ _____ _____ _____
Thousands   Hundreds   Tens   Units

In July of this year, Idaho and Wyoming were admitted as the 43rd and 44th states of the Union. Idaho was admitted on July 3 and Wyoming on July 10. To learn the year, just solve this puzzle.

- The two-digit number formed by my tens and units digits is the number of degrees in a right angle.

- The two-digit number formed by my thousands and hundreds digits has this prime factorization: 2×3^2.

- The sum of all of my digits is the same as the number of holes on many golf courses.

What year am I?

_____	_____	_____	_____
Thousands	Hundreds	Tens	Units

Utah became the 45th state in January of this year. Utah was admitted to the Union on January 4. To learn the year, just solve this puzzle.

- My units digit is equal to 3!.

- The two-digit number formed by my hundreds and tens digits is the largest prime number between 80 and 90.

- The sum of all of my digits is equal to 4!.

What year am I?

| _____ | _____ | _____ | _____ |
| Thousands | Hundreds | Tens | Units |

Oklahoma became the 46th state when it was admitted to the Union on November 16 of this year. To learn the year, just solve this puzzle.

- My units digit is the same as the number of sides on a heptagon.

- The two-digit number formed by my hundreds and tens digits is the same as the sum of the first nine even numbers (starting with 2).

- The sum of all of my digits is the median of this set of numbers: 2, 4, 8, 12, 17, 31, 32, 67, 91.

What year am I?

| _____ | _____ | _____ | _____ |
| Thousands | Hundreds | Tens | Units |

In this year, New Mexico and Arizona became the 47th and 48th states admitted to the Union. New Mexico became a state on January 6 and Arizona on February 14. To learn the year, just solve this puzzle.

- The two-digit number formed by my tens and units digits is the number of faces on a dodecahedron.

- The two-digit number formed by my hundreds and tens digits is equal to $1 + 2 + 3 + 4 + \ldots + 13$.

- The sum of all of my digits is the number of cards in each suit in a standard deck of playing cards.

What year am I?

34

_____   _____   _____   _____
Thousands   Hundreds   Tens   Units

The last two states to be admitted to the Union, Alaska (#49) and Hawaii (#50) were both admitted in the same year. Alaska was admitted on December 3 and Hawaii on August 21. To learn the year, just solve this puzzle.

• My hundreds digit and units digit are the same number—the number of sides in a nonagon.

• My tens digit is the fifth number in the Fibonacci sequence.

• If a girl had 4 shirts, 3 skirts, and 2 pair of shoes and chose 1 of each, she would have the same number of different outfits as the sum of all of my digits.

What year am I?

_____   _____   _____   _____
Thousands   Hundreds   Tens   Units

Famous Firsts

We are all fascinated by the first of anything—the first person to climb a mountain, the first minority to be elected to office, even the first telephone operator! These puzzles take us back in time to contemplate those explorers, scientists, entrepreneurs, and others who have changed history by being the first of something. By solving these puzzles, students learn the year that the "famous first" occurred.

The first women to become FBI agents completed their training in Quantico, Virginia, in this year. The new agents were Susan Roley and Joanne Pierce. Solve this puzzle to learn the year.

- My odd tens and even units digits are both prime numbers; if they were the sides of a rectangle, its area would be 14.

- The sum of my tens and units digits is equal to my hundreds digit.

- The sum of all of my digits is 1 less than 2 decades.

What year am I?

_____   _____   _____   _____
Thousands   Hundreds   Tens   Units

On January 2 of this year, the first junior high school was opened in the United States. Solve this puzzle to learn the year.

- The two-digit number formed by my tens and units digits is equal to the sum of the first four counting numbers.

- My hundreds digit is equal to 3^2.

- The sum of all of my digits is the same as the number of players each team has on a football field.

What year am I?

_____   _____   _____   _____

Thousands   Hundreds   Tens   Units

37

© 2003 J. Weston Walch, Publisher

Nellie Tayloe Ross became the first woman to be elected a governor in the United States. She took the oath of office on January 5 of this year. The state was Wyoming.

- The product of my tens and units digits is equal to the sum of my hundreds and thousands digits.

- My tens and units digits are prime numbers, although the tens digit is even and the units digit is odd.

- The two-digit number formed by my tens and units digits could be the area of a square with a perimeter of 20.

What year am I?

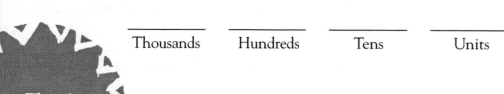

| Thousands | Hundreds | Tens | Units |

38

On January 9 of this year, Frenchman François Jean Pierre Blanchard crossed between Philadelphia and New Jersey in the first successful balloon flight in the United States. The balloon reached a height of over 5,800 feet and traveled 15 miles in 45 minutes. (How many miles per hour is this?)

- The two-digit number formed by my tens and units digits is equal to the sum of 13 + 14 +... + 17 + 18.

- The sum of my thousands digit and my prime hundreds digit is 1 less than my tens digit.

What year am I?

_____	_____	_____	_____
Thousands	Hundreds	Tens	Units

39

The first woman U.S. senator, Hattie Caraway, was elected from the state of Arkansas in this year. Solve the puzzle to learn the year.

- The two-digit number formed by my tens and units digits is equal to the 5th power of 2.

- My hundreds digit is 300% of my tens digit.

- The sum of all of my digits is equal to the sum of the first five counting numbers.

What year am I?

_____ _____ _____ _____
Thousands   Hundreds   Tens   Units

40

On January 13 of this year, Robert C. Weaver became the first African-American cabinet member when President Lyndon B. Johnson appointed him Secretary of Housing and Urban Development. Learn the year by solving this puzzle.

- The prime factors of the two-digit number formed by my hundreds and tens digits are $2^5 \times 3$.

- My units digit is equal to 3!.

- The sum of all of my digits is equal to the number of inches in $1\frac{5}{6}$ feet.

What year am I?

_____   _____   _____   _____
Thousands   Hundreds   Tens   Units

41

On January 21 of this year, the first law requiring that drivers of automobiles have licenses went into effect. To learn the year, solve this puzzle.

- The two-digit number formed by my tens and units digits is the 12th prime number and is considered normal body temperature when measured in degrees Celsius.

- My tens digit is $33\frac{1}{3}\%$ of my hundreds digit.

- The sum of all of my digits is the same as the number of triangles that make up an icosahedron.

What year am I?

_____   _____   _____   _____
Thousands   Hundreds   Tens   Units

In this year, Dr. Elizabeth Blackwell became the first woman to receive an M.D. degree. She attended the Medical Institute of Geneva, New York. Solve this puzzle to find the year.

- The number in my hundreds digit is the same as the number of vertices on a cube.

- The two-digit number formed by my tens and units digits is the same as the sum of the first seven odd numbers.

- The sum of all of my digits is the same as the number of pints in 2 gallons, 3 quarts.

What year am I?

| _____ | _____ | _____ | _____ |
| Thousands | Hundreds | Tens | Units |

43

On January 28 of this year, Louis Brandeis became the first American Jew to be appointed a United States Supreme Court justice. Solve this puzzle to learn the year.

- The two-digit number formed by my tens and units digits is equal to the 4th power of 2.

- The two-digit number formed by my thousands and hundreds digits is the largest prime number less than 20.

What year am I?

_____   _____   _____   _____
Thousands   Hundreds   Tens   Units

On January 31 of this year, the first daytime soap opera, *These Are My Children*, was broadcast from the NBC studios in Chicago. Solve this puzzle to learn the year.

- The square root of the two-digit number formed by my tens and units digits is equal to 7.

- My hundreds digit is equal to the number of inches in $\frac{3}{4}$ of a foot.

- The sum of all of my digits is the smallest prime number between 20 and 30.

What year am I?

_____   _____   _____   _____
Thousands   Hundreds   Tens   Units

45

Maggie Walker became the first African-American woman to establish and manage a bank in this year. Solve this puzzle to learn the date.

- My units digit is the square root of my hundreds digit.

- The two-digit number formed by my hundreds and tens digits is equal to the sum of the first ten even numbers.

- The sum of all of my digits is equal to the number of doughnuts in a baker's dozen.

What year am I?

———————   ———————   ———————   ———————
Thousands   Hundreds   Tens   Units

46

On February 10 of this year, Western Union delivered the first singing telegram. Solve the puzzle to learn the year.

- The two-digit number formed by my tens and units digits is equal to the sum of 1! + 2! + 3! + 4!.

- My hundreds digit is equal to $\sqrt{81}$.

- The sum of all of my digits is equal to the number of ounces in 1 pound.

What year am I?

_____   _____   _____   _____

Thousands   Hundreds   Tens   Units

47

On February 10 of this year, the familiar slogan "All the News That's Fit to Print" first appeared on the front page of *The New York Times*. The newspaper offered a prize of $100 to anyone who could come up with a better slogan in 10 words or fewer. Although thousands of submissions were received, none was judged better. It is still the newspaper's slogan. Solve the puzzle to find the year.

- My hundreds, tens, and units digits are consecutive integers.
- My hundreds digit is even; my tens digit is odd; my units digit is prime.
- The sum of all of my digits is equal to 5^2.

What year am I?

_____ _____ _____ _____
Thousands   Hundreds   Tens   Units

48

On February 20 of this year, astronaut John Glenn became the first American to orbit the earth. He circled the globe three times before landing.

- The two-digit number formed by my hundreds and tens digits is equal to the sum of the odd numbers greater than or equal to 5 and less than or equal to 19.

- My units digit is $\frac{1}{3}$ of my tens digit.

- The sum of all of my digits is equal to 2×3^2.

What year am I?

_____ _____ _____ _____

Thousands   Hundreds   Tens   Units

49

On March 2 of this year, the first school for the blind in America was established in Massachusetts. Solve this puzzle to learn the year.

- My hundreds digit is equal to 2^3 while my units digits is equal to 3^2.

- The digit in my tens place is equal to $2!$.

- The sum of all of my digits is equal to the number of equilateral triangles in an icosahedron.

What year am I?

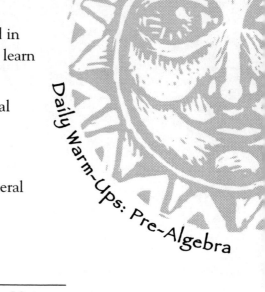

_____ _____ _____ _____
Thousands   Hundreds   Tens   Units

50

Charles Brady King drove the first automobile in Detroit on March 6 of this year. Solve this puzzle to learn the year.

- The two-digit number formed by my tens and units digits is equal to $2^5 \times 3$.

- My hundreds digit is $133\frac{1}{3}\%$ of my units digit.

- The sum of all of my digits is the same as 2 dozen.

What year am I?

_____   _____   _____   _____
Thousands   Hundreds   Tens   Units

51

© 2003 J. Weston Walch, Publisher

On March 12 of this year, the first parachute jump from an airplane in the United States was made. (What do you think the statement, "Minds are like parachutes, they only function when they're open" means?)

- The two-digit number formed by my tens and units digits could be the perimeter of a rectangle with a length of 4 and a width of 2.

- The sum of all of my digits is the same as a baker's dozen.

What year am I?

| _____ | _____ | _____ | _____ |
| Thousands | Hundreds | Tens | Units |

52

On March 16 of this year, the first African-American newspaper in the United States was published in New York City. It was called *Freedom's Journal*. The day is now celebrated as Black Press Day in New York.

- My hundreds digit is equal to my tens digit raised to the 3rd power.

- My units digit is 1 less than my hundreds digit.

- The sum of all of my digits is equal to 2×3^2.

What year am I?

Thousands	Hundreds	Tens	Units

53

On March 19 of this year, the swallows first returned to the old mission of San Juan Capistrano, California. To learn the year, solve this puzzle.

- The two-digit number formed by my hundreds and tens digits is equal to the product of the 4th and 5th prime numbers.

- My units digit is a perfect number.

- The sum of all of my digits is the same as the sum of the first six counting numbers.

What year am I?

_____   _____   _____   _____
Thousands   Hundreds   Tens   Units

54

On March 22 of this year, the first women's collegiate basketball game was played at Smith College in Northampton, Massachusetts. Sandra Berenson, "the Mother of Women's Basketball," supervised the game.

- My hundreds and tens digits are consecutive integers with a sum of 17 and a product of 72.

- My units digit is $\frac{1}{3}$ of my tens digit.

- The sum of all of my digits is equal to 3 less than 2 dozen.

What year am I?

_____   _____   _____   _____

Thousands   Hundreds   Tens   Units

55

Henry "Hank" Aaron hit the 715th home run of his career and broke Babe Ruth's record on April 8 of this year. He finished his career with a total of 755 home runs. Solve this puzzle to learn the year.

- The prime factorization of the two-digit number formed by my tens and units digits is equal to the product of 2 and 37.

- The sum of all of my digits is equal to 21.

What year am I?

| _____ | _____ | _____ | _____ |
| Thousands | Hundreds | Tens | Units |

56

On April 10 of this year, the Brooklyn Dodgers recruited Jackie Robinson. He was the first African American to play on a major league baseball club. Solve this puzzle to learn the year.

- The sum of my even tens digit and my prime units digit is 1 greater than the sum of my odd thousands and hundreds digits.

- The sum of my digits is equal to 1 more than the number of years in 2 decades.

What year am I?

_____   _____   _____   _____
Thousands   Hundreds   Tens   Units

57

On April 19 of this year, the Battle of Lexington began the Revolutionary War. The first gunfire was called "the shot heard 'round the world." To learn the year, solve this puzzle.

- The two-digit number formed by my tens and units digits is the sum of the integers greater than or equal to 3 and less than or equal to 12.

- My hundreds digit is equal to $3^2 - 2^1$.

- The sum of all of my digits is equal to $\dfrac{7 + (6 \times 4) + 9}{2}$.

What year am I?

_____	_____	_____	_____
Thousands	Hundreds	Tens	Units

58

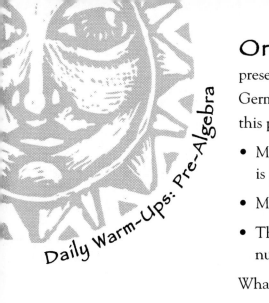

On April 25 of this year, the first Seeing Eye dog was presented in America. It has been found that purebred female German shepherds make the most effective Seeing Eye dogs. Solve this puzzle to find the year.

- My tens and units digits are consecutive integers whose product is $\frac{2}{3}$ of my hundreds digit.

- My units digit is $\frac{1}{3}$ of my hundreds digit.

- The sum of my digits is equal to the sum of the first five counting numbers.

What year am I?

_____ _____ _____ _____
Thousands   Hundreds   Tens   Units

Gwendolyn Brooks became the first African-American woman to win the Pulitzer Prize on May 5th of this year. Solve this puzzle to learn the year.

- The two-digit number formed by my tens and units digits is equal to the value of the Roman numeral L.

- My hundreds digit is equal to $\sqrt{81}$.

- The sum of all of my digits is $\frac{1}{2}$ $(8 \times 4 - 2)$.

What year am I?

_____ _____ _____ _____

Thousands   Hundreds   Tens   Units

Hernando de Soto became the first European to reach the Mississippi River on May 8 of this year. Solve this puzzle to learn the year.

- The two-digit number formed by my hundreds and tens digits is equal to 2×3^3.

- Both my thousands and units digits are the multiplicative identity element.

What year am I?

_____	_____	_____	_____
Thousands	Hundreds	Tens	Units

61

© 2003 J. Weston Walch, Publisher

The first ocean-to-ocean railroad was

completed on May 10 of this year. The Union Pacific and Central Pacific railways were linked at Promontory Point, Utah, by a golden spike. Learn the year by solving this puzzle.

- My tens digit is 75% of my hundreds digit.
- My tens digit is $66\frac{2}{3}$% of my units digit.
- The sum of all of my digits is equal to 4!.

What year am I?

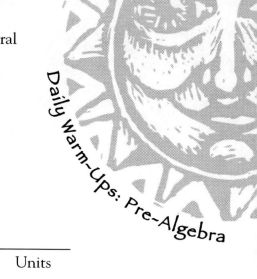

_____ _____ _____ _____

Thousands Hundreds Tens Units

62

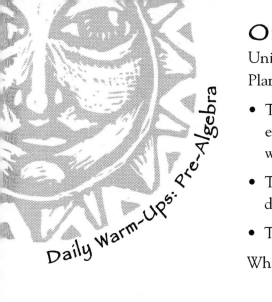

On May 10 of this year, the first planetarium in the United States opened in the city of Chicago. It is called the Adler Planetarium. Solve this puzzle to learn the year.

- The two-digit number formed by my tens and units digits is equal to the sum of the first five even numbers (beginning with 2).

- The two-digit number formed by my thousands and hundreds digits is the eighth prime number.

- The sum of all of my digits is the sixth prime number.

What year am I?

| _____ | _____ | _____ | _____ |
| Thousands | Hundreds | Tens | Units |

63

On June 18 of this year, Sally Ride became the first American woman in space. She functioned as the mission specialist on a six-day flight of the space shuttle *Challenger*. Solve this puzzle to learn the year.

- The two-digit number formed by my hundreds and tens digits is equal to 2×7^2.

- My units digit is the first odd prime number.

- The sum of all of my digits is equal to $-3 + 2^3 \times 3$.

What year am I?

| Thousands | Hundreds | Tens | Units |

© 2003 J. Weston Walch, Publisher

On July 20 of this year, astronaut Neil Armstrong became the first person to land on the moon. When he stepped on the moon, he said, "That's one small step for man, one giant leap for mankind." To learn the year, solve this puzzle.

- My tens digit is a perfect number.

- My hundreds and units digits could be the sides of a square with an area of 81 and a perimeter of 36.

- The sum of all of my digits is equal to $3^2 + 4^2$.

What year am I?

| _____ | _____ | _____ | _____ |
| Thousands | Hundreds | Tens | Units |

On August 17 of this year, three Americans—Max Anderson, Ben Abruzzo, and Larry Newman—became the first to cross the Atlantic in a hot-air balloon. They traveled 3,200 miles in a little over 137 hours. Learn the year by solving this puzzle.

- My hundreds, tens, and units digits are consecutive integers that are out of order; their mean is 8.

- My hundreds digit is a square number, my tens digit is prime, and my units digit is cubic.

- The sum of all of my digits is equal to $\sqrt{125}$.

What year am I?

_____   _____   _____   _____
Thousands   Hundreds   Tens   Units

Daily Warm-Ups: Pre-Algebra

On August 30 of this year, Esther Cleveland became the first baby to be born to the wife of a president in the White House. Learn the year of this famous first by solving this puzzle.

- My units digit is $\frac{1}{3}$ of my tens digit.

- The sum of my thousands and hundreds digits is equal to my tens digit.

- The sum of all of my digits is equal to the sum of the first six counting numbers.

What year am I?

| _____ | _____ | _____ | _____ |
| Thousands | Hundreds | Tens | Units |

On September 25 of this year, the first and only edition of *Public Occurrences Both Foreign and Domestick* was published in Boston. Although this was the first American newspaper, the authorities considered it to be "offensive," and it was suppressed immediately. Solve this puzzle to learn the year.

- My hundreds digit is 3 less than my tens digit.

- The two-digit number formed by my tens and units digits is equal to the sum of the even integers greater than or equal to 2 and less than or equal to 18.

- The sum of all of my digits is equal to 4^2.

What year am I?

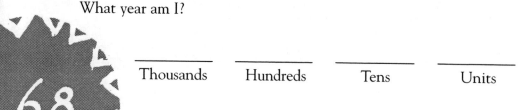

Thousands ___ Hundreds ___ Tens ___ Units

68

On October 2 of this year, Thurgood Marshall was sworn in as the first African-American associate justice of the United States Supreme Court. Learn the year by solving this puzzle.

- My tens digit is equal to 3!.
- My prime units digit is 2 less than my odd hundreds digit.
- The sum of all of my digits is equal to $16 + 3 \times 4 - \sqrt{25}$.

What year am I?

_____   _____   _____   _____
Thousands   Hundreds   Tens   Units

69

© 2003 J. Weston Walch, Publisher

Discoveries, Inventions, and Notable Accomplishments

Looking at a time line of the history of humankind, one might consider the changes that have come to pass during the 20th century as nothing less than amazing! My grandmother was born before there were automobiles, my mother was born before there were airplanes, and I was born before there were jet airplanes or men on the moon! In three generations, we have gone from the horse and buggy era to an International Space Station that orbits the earth. This section looks at those discoveries, inventions, and other notable accomplishments that have changed our world. When students solve each puzzle, they learn the year of each of the accomplishments.

On January 5 of this year, German physicist Wilhelm Roentgen announced the discovery of the X ray. Solve this puzzle to learn the year.

- The two-digit number formed by my thousands and hundreds digits is equal to $2 \times \sqrt{81}$.

- The two-digit number formed by my tens and units digits is the product of the third and eighth prime numbers.

- The sum of all of my digits is 2 less than 5^2.

What year am I?

| _____ | _____ | _____ | _____ |
| Thousands | Hundreds | Tens | Units |

70

On January 5 of this year, Lizzie Magie received a patent for her board game called "The Landlord's Game." It is very similar to "Monopoly" except all the properties in Magie's game were rented, not purchased. To learn the year Magie received her patent, solve this puzzle.

- My tens digit is the additive identity element.

- My hundreds and units digits are both square numbers, but one is odd and the other is even.

- The two-digit number formed by my thousands and hundreds digits is the eighth prime number.

What year am I?

_____   _____   _____   _____
Thousands   Hundreds   Tens   Units

71

On January 13 of this year, the accordion was patented. This is a musical instrument with keys, metal reeds, and bellows. By fingering the keys, air is forced through the reeds by closing and opening the bellows. Solve this puzzle to learn the year.

- The two-digit number formed by my tens and units digits is divisible by 1, 2, 3, 6, 9, 18, 27, and itself.

- The sum of my thousands and hundreds digits is equal to the sum of my tens and units digits.

- The sum of all of my digits is equal to the number of inches in $1\frac{1}{2}$ feet.

What year am I?

72

_____   _____   _____   _____
Thousands   Hundreds   Tens   Units

On January 14 of this year, the Pentagon building in Washington, D.C., was completed. It is the five-sided building that houses the Department of Defense. To learn the year, solve this puzzle.

- My hundreds and tens digits are both square numbers, but one is odd and the other is even.

- My units digit is the missing number in this sequence: 1, 1, 2, _, 5, 8, 13 . . .

- The sum of all of my digits is equal to the seventh prime number.

What year am I?

| _____ | _____ | _____ | _____ |
| Thousands | Hundreds | Tens | Units |

© 2003 J. Weston Walch, Publisher

On January 24 of this year, a gold nugget was found at the site of a sawmill owned by John Sutter near Colona, California. This discovery started the California Gold Rush. Learn the year by solving this puzzle.

- My even hundreds, tens, and units digits form a palindrome whose sum is 20.

- My hundreds digit is 200% of my tens digit.

- The sum of all of my digits is equal to the number of days in three weeks.

What year am I?

74

_____ _____ _____ _____
Thousands   Hundreds   Tens   Units

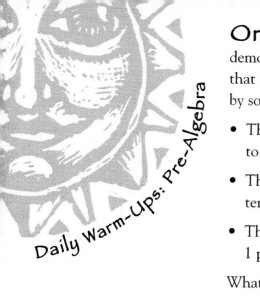

On February 12 of this year, Alexander Graham Bell demonstrated how his invention, the telephone, worked with a line that ran between Boston and Salem, Massachusetts. Learn the year by solving this puzzle.

- The two-digit number formed by my tens and units digits is equal to the product of the fourth and fifth prime numbers.

- The sum of my thousands and hundreds digit is 2 greater than my tens digit.

- The sum of all of my digits is equal to the number of ounces in 1 pound 7 ounces.

What year am I?

_____   _____   _____   _____
Thousands   Hundreds   Tens   Units

On February 16 of this year, the tomb of King Tutankhamen was opened by archaeologists. It had been sealed for more than 3,000 years. Learn the year of this discovery by solving this puzzle.

- The square of my odd units digit is equal to my hundreds digit.

- My tens digit is the same as the number of points needed to form a line.

- The sum of all of my digits is equal to the sum of the first five counting numbers.

What year am I?

_____ _____ _____ _____
Thousands   Hundreds   Tens   Units

76

On February 19 of this year, Thomas Edison received a patent for his invention, the phonograph. It was called the "first talking machine." Recordings at that time were made on wax cylinders. Solve this puzzle to learn the year.

- My hundreds, tens, and units digits form a palindrome with a sum of 23.

- My entire date is divisible by 2, 3, and 6.

- The sum of my digits is equal to the sum of the odd integers that are greater than or equal to 3 and less than or equal to 9.

What year am I?

_____ _____ _____ _____

Thousands   Hundreds   Tens   Units

On March 18 of this year, Schick marketed the first electric razor. Learn the year by solving this puzzle.

- My hundreds digit is 300% of my tens digit.
- My units digit is the multiplicative identity element.
- The sum of all of my digits is equal to $-4 + 6 + 3 \times 2^2$.

What year am I?

_____ _____ _____ _____
Thousands   Hundreds   Tens   Units

On March 17 of this year, paper money became legal tender by an act of Congress. The denominations were $5, $10, and $20. To learn when this money was created, solve this puzzle.

- My hundreds and tens digits could be the sides of a rectangle that has an area of 48 square units and a perimeter of 28 units.

- My units digit is $\frac{1}{4}$ of my hundreds digit and $\frac{1}{3}$ of my tens digit.

- The sum of all of my digits is the same as the seventh prime number.

What year am I?

_____   _____   _____   _____

Thousands   Hundreds   Tens   Units

79

On March 26 of this year, Dr. Jonas Salk introduced the polio vaccine in the United States. To learn the year of this medical discovery that helped wipe out a terrible disease, solve this puzzle.

- The product of my tens and units digits is 15.
- My hundreds digit is 300% of my units digit.
- The sum of all of my digits is equal to 2×3^2.

What year am I?

_____   _____   _____   _____
Thousands   Hundreds   Tens   Units

On April 6 of this year, Teflon® was invented by Dr. Roy Plunkett at the DuPont research laboratories. The surface of Teflon is so slippery that virtually nothing sticks to or is absorbed by it. To learn the year of this discovery, solve this puzzle.

- The two-digit number formed by my tens and units digits is equal to the sum of the whole numbers greater than or equal to 8 and less than or equal to 11.

- My tens digit is $33\frac{1}{3}\%$ of my hundreds digit.

- The sum of all of my digits is equal to the sum of the first six counting numbers.

What year am I?

Thousands	Hundreds	Tens	Units
_____	_____	_____	_____

81

© 2003 J. Weston Walch, Publisher

On April 20 of this year, Marie and Pierre Curie isolated the element radium. This husband-and-wife team was awarded the Nobel Prize for their work. Later, Marie Curie won a second Nobel Prize for her work in chemistry, making her the first person to win this prize twice. Even more amazing, Marie Curie's daughter, Irène Joliot-Curie, was also awarded a Nobel Prize for Chemistry. Solve this puzzle to learn the year radium was isolated.

- My units digit raised to the 3rd power would be 1 less than my hundreds digit.

- My tens digit is the additive identity element.

- The sum of all of my digits is equal to $2^2 \times 3$.

    What year am I?

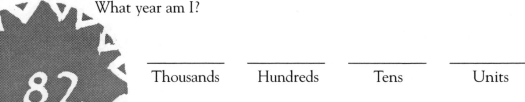

82

Thousands	Hundreds	Tens	Units
_____	_____	_____	_____

In this year, Ruth Handler, the co-founder of Mattel toys, introduced her invention, the Barbie doll. She named the doll after her daughter, Barbara. The first doll sold for only $3.00. One of these original dolls, in mint condition, has sold for up to $10,000.00. To learn the year of this invention, just solve this puzzle.

- My hundreds, tens, and units digits form a palindrome whose sum is 23.

- My tens digit is 50% of the sum of my thousands and hundreds digits.

- The sum of all of my digits is equal to 4!.

What year am I?

Thousands	Hundreds	Tens	Units
_____	_____	_____	_____

83

Unhappy with cloth diapers that leaked and had to be washed, Marion Donovan invented the convenient disposable diaper in this year. When companies thought her product would be too expensive to produce, she went into business for herself. A few years later, she sold her business for $1 million. Solve this puzzle to learn the year the disposable diaper was invented.

- The two-digit number formed by my tens and units digits has the same value as the Roman numeral L.

- My hundreds digit is 1 less than twice my tens digit.

- The sum of all of my digits is equal to the sum of the first five counting numbers.

What year am I?

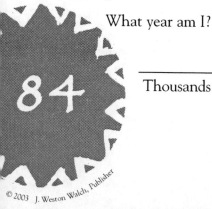

_____   _____   _____   _____
Thousands   Hundreds   Tens   Units

Noah McVicker and Joseph McVicker invented Play-Doh in this year. It received U.S. Patent No. 3,167,440 nine years later. Over 700 million pounds of Play-Doh have been sold. To learn the year this toy was invented, solve this puzzle.

- The two-digit number formed by my tens and units digits is equal to the sum of the even numbers greater than or equal to 2 and less than 16.

- The sum of all of my digits is equal to the number of feet in 7 yards.

What year am I?

_____ _____ _____ _____
Thousands   Hundreds   Tens   Units

© 2003 J. Weston Walch, Publisher

On April 26 of this year, Sarah Boone, one of the first African-American women to receive a patent, invented an improved ironing board. To learn the year of this invention, solve this puzzle.

- My hundreds and tens digits are consecutive integers with a sum of 17 and a product of 72.

- My units digit is 25% of my hundreds digit.

- The sum of all of my digits is the same as the number of triangular faces in an icosahedron.

What year am I?

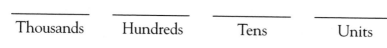

<u>   </u>   <u>   </u>   <u>   </u>   <u>   </u>

Thousands   Hundreds   Tens   Units

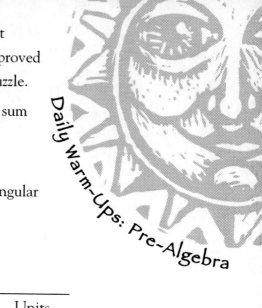

86

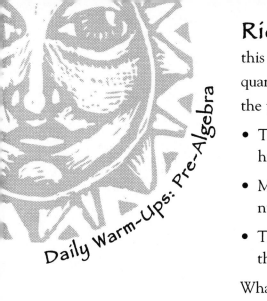

Richard and Betty James invented the Slinky® in this year. The Slinky is made from 80 feet of wire, and over a quarter of a billion Slinkys have been sold worldwide. To learn the year this toy was invented, solve this puzzle.

- The sum of my consecutive tens and units digits is equal to my hundreds digit; my date is divisible by 5.

- My hundreds digit is equal to the sum of the first three odd numbers.

- The sum of all of my digits is the largest prime number less than 20.

What year am I?

_____ _____ _____ _____
Thousands   Hundreds   Tens   Units

87

On May 19 of this year, the Simplon Tunnel connecting Switzerland and Italy was officially opened. To learn the year, solve this puzzle.

- My units digit is equal to 3!.

- The two-digit number formed by my hundreds and tens digits is equal to the sum of the even numbers from 2 to 18.

- The sum of all of my digits is equal to the number of quarts in 4 gallons.

What year am I?

_____   _____   _____   _____
Thousands   Hundreds   Tens   Units

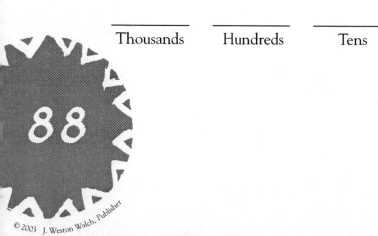

88

On May 20 of this year, 25-year-old aviator Captain Charles Lindbergh departed from a rainy field in New York to begin the first solitary flight across the Atlantic. His monoplane was called the *Spirit of St. Louis*; the 3,600-mile flight took almost 39 hours. Solve this puzzle to learn the year.

- Three times my hundreds digit is equal to the two-digit number formed by my tens and units digits.

- The sum of my tens and units digits is equal to my hundreds digit.

- The sum of all of my digits is the eighth prime number.

What year am I?

_____ _____ _____ _____
Thousands   Hundreds   Tens   Units

89

On May 24 of this year, the Brooklyn Bridge opened in New York City. It took 14 years to construct and cost over $16 million; it crosses the East River, connecting Brooklyn and Manhattan. This steel suspension bridge was designed by John Roebling and has a span of 1,595 feet. Solve this puzzle to learn the year.

- My hundreds and tens digits could be the sides of a square with a perimeter of 32 units.

- My units digit is the smallest odd prime number.

- The sum of all of my digits is equal to $2^2 \times 5$.

What year am I?

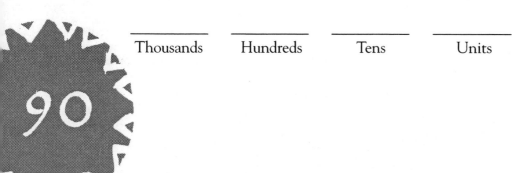

_____   _____   _____   _____

Thousands   Hundreds   Tens   Units

90

Bette Nesmith Graham, a secretary in Dallas, invented liquid paper in this year. She mixed up the first batch in her home blender to correct her typing errors. She called it "Mistake Out" and started the business out of the basement of her home. Thirty-four years later, she sold the business for $47.5 million. Solve this puzzle to learn the year Graham invented "Mistake Out."

- My tens and units digits are consecutive integers with a sum of 11 and a product of 30.

- My units digit is $\frac{2}{3}$ of my hundreds digit.

- The sum of all of my digits is equal to the total number of spots on one die.

What year am I?

_____ _____ _____ _____

Thousands   Hundreds   Tens   Units

91

On May 28 of this year, the Dionne quintuplets (Marie, Cecile, Yvonne, Emilie, and Annette) were born in Ontario, Canada. They were the first quintuplets known to survive for more than a few hours after birth. Learn the year they were born by solving this puzzle.

- My tens digit is $\frac{1}{3}$ of my hundreds digit.

- My units digit is equal to $\sqrt{16}$.

- The sum of all of my digits is the seventh prime number.

What year am I?

_____   _____   _____   _____
Thousands   Hundreds   Tens   Units

92

On June 3 of this year, the famous poem, "Casey at the Bat" was first printed in the *San Francisco Examiner*. It was written by Ernest Thayer. Solve this puzzle to learn the year.

- My hundreds, tens, and units digits could be the sides of a cube with a volume of 512 cubic units.

- My date is divisible by 2.

- The sum of all of my digits is equal to $3^2 + 4^2$.

What year am I?

_____   _____   _____   _____
Thousands   Hundreds   Tens   Units

On June 8 of this year, John McGaffrey patented the first suction-type vacuum cleaner. To learn the year, solve this puzzle.

- My hundreds and tens digits are even; my tens digit is $\frac{3}{4}$ of my hundreds digit.

- My units digit is odd; it is 150% of my tens digit.

- The sum of all of my digits is equal to two dozen.

What year am I?

_____   _____   _____   _____
Thousands   Hundreds   Tens   Units

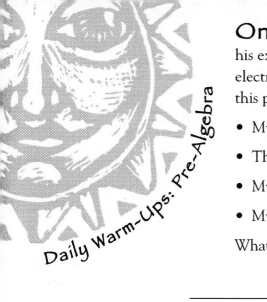

On June 15 of this year, Benjamin Franklin conducted his experiment with a kite and proved that lightning contained electricity. To learn the year of this *electrifying* discovery, solve this puzzle.

- My date is divisible by 2.
- The sum of my tens and units digits is equal to my hundreds digit.
- My hundreds digit is the fourth prime number.
- My units digit is the only even prime number.

What year am I?

_____   _____   _____   _____

Thousands   Hundreds   Tens   Units

95

© 2003 J. Weston Walch, Publisher

On June 21 of this year, the wheat reaper was patented. Before this invention, only about 3 acres of wheat could be harvested each day; with the reaper, about 15 acres could be harvested. (What percent of increase does this represent?) Learn the year of this patent by solving this problem.

- My tens and units digits are consecutive integers with a sum of 7 and a product of 12.

- My date is divisible by 2.

- My hundreds digit is the same as the number of sides on an octagon.

- The sum of all of my digits is equal to the number of ounces in 1 pound.

  What year am I?

_____ _____ _____ _____
Thousands   Hundreds   Tens   Units

Scientist Patsy Sherman had a very fortunate accident in this year. While trying to develop a rubber hose to use in jet aircraft, she discovered a substance that repelled both water and oil. This substance is now known as Scotchgard Fabric Protector. Learn the year this fortunate event occurred by solving this puzzle.

- The two-digit number formed by my tens and units digits is the sum of the integers greater than or equal to 3 and less than or equal to 10.

- My hundreds digit is $4\frac{1}{2}$ the size of my units digit.

- The sum of all of my digits is the seventh prime number.

What year am I?

| _____ | _____ | _____ | _____ |
| Thousands | Hundreds | Tens | Units |

97

Joseph Woodland and Bernard Silver received a patent for their invention, the bar code, in this year. A bar code is used for automatic identification and data collection. It eventually became what we now call a UPC symbol (Universal Product Code). Twenty-four years after it was invented, the first UPC scanner was used in a store. Solve this puzzle to learn the year of this invention.

- The two-digit number formed by my tens and units digits is equal to 3 + 4 + . . . + 9 + 10.

- My hundreds digit is equal to the sum of the first three odd integers.

- The sum of all of my digits is the seventh prime number.

      What year am I?

<u>   </u>   <u>   </u>   <u>   </u>   <u>   </u>

  Thousands   Hundreds   Tens   Units

98

On November 13 of this year, the Holland Tunnel was completed in New York City. Running under the Hudson River, it connects Manhattan with Jersey City, New Jersey. Solve this puzzle to learn the year.

- The two-digit number formed by my tens and units digits is 3 times my hundreds digit.

- My units digit is the largest single-digit prime.

- The sum of all of my digits is equal to the number of inches in $1\frac{7}{12}$ feet.

What year am I?

| _____ | _____ | _____ | _____ |
| Thousands | Hundreds | Tens | Units |

99

On November 2 of this year, the largest airplane ever made took its one and only flight over Long Beach Harbor in California. The 200-ton plywood craft, which cost $25 million to build, was named the *Spruce Goose*. Learn the year of this flight by solving this puzzle.

- My units digit is equal to $2^3 - 1$; my tens digit is 2^2.

- The two-digit number formed by my thousands and hundreds digits is the eighth prime number.

- The sum of all of my digits is the missing number in this sequence: 1, 1, 2, 3, 5, 8, 13, ____, 34, 55 . . .

What year am I?

_____	_____	_____	_____
Thousands	Hundreds	Tens	Units

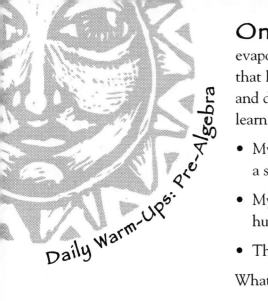

On November 25 of this year, the patent application for evaporated milk was submitted. Evaporated milk is unsweetened milk that has been thickened by partial evaporation; it is packaged in cans and does not need refrigeration until it is opened. Solve this puzzle to learn the year this patent was submitted.

- My hundreds and tens digits are the same; if they were the sides of a square, its perimeter would be 32.

- My units digit raised to the 3rd power is equal to the product of my hundreds and tens digits.

- The sum of all of my digits is equal to $4^2 + \sqrt{16} + \sqrt{1}$.

What year am I?

_____   _____   _____   _____
Thousands   Hundreds   Tens   Units

On December 4 of this year, a painting in the Metropolitan Museum of Art in New York City was found to have been hung upside down. It had been in this embarrassing position for 47 days. Solve this puzzle to learn the year.

- My tens digit is $\frac{2}{3}$ of my hundreds digit.

- Both my thousands and units digits are the multiplicative identity.

- The sum of all of my digits is equal to the seventh prime number.

What year am I?

_____   _____   _____   _____
Thousands   Hundreds   Tens   Units

On December 20 of this year, the Louisiana Purchase took place. In one of the greatest real estate deals in history, the United States purchased more than a million square miles of land from France for about $20 per square mile. Learn the year of this purchase by solving this puzzle.

- My hundreds digit is a cubic number and is 1 less than 3 times my units digit.

- The two-digit number formed by my hundreds and tens digits is equal to $2^4 \times 5$.

- The sum of all of my digits is equal to one dozen.

What year am I?

_____	_____	_____	_____
Thousands	Hundreds	Tens	Units

103

© 2003 J. Weston Walch, Publisher

Happy Birthday to You . . .

The puzzles in this section all relate to the birthdays of famous (and sometimes not so famous) people. Because many of these people will be familiar to students, they may analyze the clues by focusing on the *historical context* rather than on the *isolated* mathematics problem. It is possible you will hear, "That answer doesn't make sense. This person had to be born in the 1900s!" This is a good thing! When students start to look at mathematics as a "sense-making experience," they begin to understand the power of mathematics, and they evolve into real problem-solvers.

On January 1 of this year, American patriot Paul Revere was born in Boston, Massachusetts. He is best known through Longfellow's poem "The Midnight Ride of Paul Revere." Solve this puzzle to learn the year.

- My tens and units digits are primes whose sum is equal to the sum of my thousands and hundreds digits.

- My date is divisible by 5.

- The sum of all of my digits is equal to the sum of the first four odd integers.

What year am I?

104

_____ _____ _____ _____
Thousands   Hundreds   Tens   Units

On January 3 of this year, author J.R.R. Tolkien was born in South Africa. His best-known works are *The Hobbit* and *The Lord of the Rings* trilogy. To learn the year this creative man was born, solve this puzzle.

- My units digit is 25% of my hundreds digit.

- My hundreds and tens digits are consecutive integers with a product of 72.

- The sum of all of my digits is the area of this triangle:

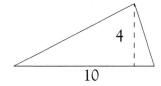

What year am I?

_____   _____   _____   _____
Thousands   Hundreds   Tens   Units

105

On January 15 of this year, Dr. Martin Luther King, Jr., was born. Dr. King was America's most famous civil rights leader, and he received the Nobel Peace Prize for his important work. To learn the year of his birth, just solve this problem.

- The two-digit number formed by my tens and units digits is a prime number and is 10 greater than the two-digit number formed by my thousands and hundreds digits.

- The sum of all of my digits is equal to the product of the second and fourth prime numbers.

What year am I?

_____   _____   _____   _____
Thousands   Hundreds   Tens   Units

On January 18 of this year, A.A. Milne was born in London, England. He is best known for the children's books *Winnie the Pooh* and *The House at Pooh Corner*. To learn the year he was born, solve this puzzle.

- My hundreds and tens digits are the same cubic number.
- My units digit is equal to $\sqrt[3]{8}$.
- The sum of all of my digits is equal to $2 + 16 \div 2 + 3 \times 3$.

What year am I?

Thousands	Hundreds	Tens	Units

107

On January 24 of this year, Native American ballerina Maria Tallchief was born. To learn the year of her birth, solve this puzzle.

- The two-digit number formed by my tens and units digits is equal to 5^2.

- The sum of my thousands and hundreds digits is equal to the product of my tens and units digits.

- The sum of all of my digits is 4 less than the number of days in three weeks.

What year am I?

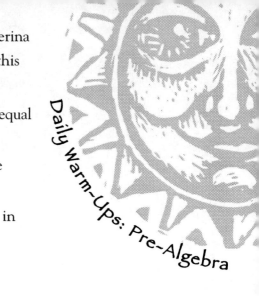

| _____ | _____ | _____ | _____ |
| Thousands | Hundreds | Tens | Units |

On January 27 of this year, composer, Wolfgang Amadeus Mozart was born in Salzburg, Austria. He began performing at the age of 3 and composed his first piece at the age of 5. He died at the age of 35. To learn the year he was born, solve this puzzle.

- My tens and units digits are consecutive integers whose sum is 11.
- My hundreds digit is the largest single-digit prime.
- My date is even.
- The sum of all of my digits is the same as the number of ounces in $1\frac{3}{16}$ pounds.

What year am I?

_____ _____ _____ _____
Thousands   Hundreds   Tens   Units

109

On February 5 of this year, African-American baseball player Hank Aaron was born in Mobile, Alabama. Aaron topped Babe Ruth's batting average and had a career total of 755 home runs. Solve this puzzle to learn the year this baseball great was born.

- My tens digit is $33\frac{1}{3}\%$ of my hundreds digit.

- The sum of my tens and units digits is 7.

- The sum of all of my digits is equal to $2 + \sqrt{4} + 3^2 + 2^2$.

What year am I?

_____   _____   _____   _____
Thousands   Hundreds   Tens   Units

110

On February 7 of this year, Laura Ingalls Wilder was born. The author of the *Little House* series, she did not begin writing books until she was 65 years old.

- My hundreds, tens, and units digits are consecutive integers but they are out of order.

- My tens digit is $\frac{3}{4}$ of my hundreds digit; my units digit is prime.

- The sum of all of my digits is equal to the sum of the integers greater than or equal to 4 and less than or equal to 7.

What year am I?

_____   _____   _____   _____
Thousands   Hundreds   Tens   Units

111

On February 13 of this year, American artist Grant Wood was born in Anamosa, Iowa. His most famous painting is called *American Gothic*. (Can you describe this painting?)

- My units digit is $\frac{1}{4}$ of my hundreds digit.

- The sum of my hundreds and thousands digits is equal to my tens digit.

- My hundreds and tens digits could be the sides of a rectangle with an area of 72 square units and a perimeter of 34 units.

What year am I?

_____   _____   _____   _____
Thousands   Hundreds   Tens   Units

112

On February 15 of this year, astronomer Galileo was born in Pisa, Italy. Galileo is credited with introducing the "scientific method" and was ridiculed for exploring unpopular scientific theories. He said, "We cannot discover new oceans unless we have the courage to lose sight of the shore."

- The two-digit number formed by my thousands and hundreds digits is equal to the sum of the first five counting numbers.

- The two-digit number formed by my tens and units digits is equal to 2^6.

- The sum of all of my digits is equal to $48 \div 2 - 2^3$.

What year am I?

_____   _____   _____   _____

Thousands   Hundreds   Tens   Units

113

© 2003 J. Weston Walch, Publisher

On February 15 of this year, Susan B. Anthony was born. A leader of the women's suffrage movement, she was arrested and fined for voting, which at the time was illegal for women. She was the first woman to appear on a United States coin. To learn the year she was born, solve this puzzle.

- My date is divisible by 2, 5, and 10.

- If my tens and hundreds digits were the sides of a rectangle, its area would be 16 square units and its perimeter would be 20 units.

- The sum of all of my digits is the fifth prime number.

What year am I?

Daily Warm-Ups: Pre-Algebra

114

_____   _____   _____   _____
Thousands   Hundreds   Tens   Units

On February 18 of this year, Russian-born author and humorist Shalom Aleichem (the pen name for Solomon Rabinowitz) was born in the Ukraine. He was affectionately known in the United States as the "Jewish Mark Twain." Solve this puzzle to learn his birth year.

- The two-digit number formed by my tens and units digits is the largest prime number less than 60.

- The two-digit number formed by my thousands and hundreds digits is equal to 2×3^2.

- The sum of all of my digits is equal to $4! - 1$.

What year am I?

_____   _____   _____   _____
Thousands   Hundreds   Tens   Units

115

On February 19 of this year, Polish astronomer Nicolaus Copernicus was born. He revolutionized scientific thought, arguing that the sun was at the center of our planetary system and that the earth revolved around it.

- The two-digit number formed by my thousands and hundreds digits is equal to the sum of the first three square numbers.

- My tens digit is 3 greater than my hundreds digit; my units digit is equal to the difference between my thousands and hundreds digits.

- The sum of all of my digits is the sum of the first five counting numbers.

What year am I?

116

_____ _____ _____ _____
Thousands   Hundreds   Tens   Units

On February 25 of this year, French Impressionist

painter Pierre-Auguste Renoir was born in Limoges, France. Renoir is noted for his delicate use of color and light. To learn the year he was born, solve this puzzle.

- My tens digit is 50% of my hundreds digit.

- My thousands and units digits are the multiplicative identity.

- The sum of all of my digits is one less than the next number in this pattern: 1, 3, 6, 10, ___.

What year am I?

| _____ | _____ | _____ | _____ |
| Thousands | Hundreds | Tens | Units |

On March 2 of this year, author and illustrator Dr. Seuss (Theodor Seuss Geisel) was born in Springfield, Massachusetts. (Can you name some of his popular books?)

- Both my hundreds and units digits are square numbers, but my hundreds digit is odd and my units digit is even.

- My tens digit is 1 less than my thousands digit.

- The sum of all of my digits is equal to the number of pints in 7 quarts.

What year am I?

Daily Warm-Ups: Pre-Algebra

_____ _____ _____ _____
Thousands   Hundreds   Tens   Units

On March 6 of this year, painter and sculptor Michelangelo (Buonarroti) was born in the town of Caprese, Tuscany. He is famous for his painting of the Sistene Chapel in Rome and his statue of David (which is located in Florence.) Solve this puzzle to learn his birth year.

- The sum of my thousands and hundreds digits is equal to my units digit.

- My date is divisible by 5 but not by 2.

- The two-digit number formed by my tens and units digits is equal to the sum of the integers 3, 4, 5, . . . 12.

What year am I?

_____   _____   _____   _____
Thousands   Hundreds   Tens   Units

119

On March 7 of this year, American naturalist Luther Burbank was born in Lancaster, Massachusetts. He is the creator and developer of many new varieties of flowers, fruits, vegetables, and trees. His birthday is celebrated as Bird and Arbor Day. Solve this puzzle to learn the day Burbank was born.

- The two-digit number formed by my tens and units digits is the sum of the odd numbers greater than or equal to 1 and less than or equal to 13.

- My hundreds digit is 200% of my tens digit.

- The sum of all of my digits is a multiple of 11.

What year am I?

120

_____ _____ _____ _____
   Thousands   Hundreds   Tens   Units

On March 14 of this year, Albert Einstein was born in Ulm, Germany. A theoretical physicist, he is best known for the theory of relativity. He immigrated to the United States and won the Nobel Prize for his work. Learn the year Einstein was born by solving this puzzle.

- My thousands and hundreds digits equal the sum of the first three multiples of 3.

- My tens digit is 2 less than my units digit and equal to the difference between my thousands and hundreds digits.

- The sum of my digits is equal to $\sqrt{125}$.

What year am I?

_____ _____ _____ _____
Thousands   Hundreds   Tens   Units

121

On March 21 of this year, Mexican resistance hero Pablo Juárez was born to Zapotec Indian parents. Orphaned at an early age, he became the symbol of Mexican resistance to foreign intervention. Learn the year Juárez was born by solving this puzzle.

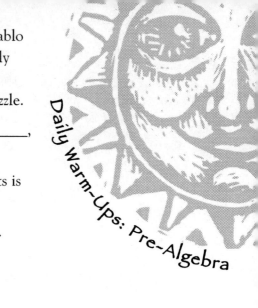

- My units digit is the missing number in this sequence: 1, 3, ____, 10, 15, 21.

- The two-digit number formed by my hundreds and tens digits is equal to $3^4 - 1$.

- The sum of all of my digits is equal to the fifth multiple of 3.

What year am I?

| _____ | _____ | _____ | _____ |
| Thousands | Hundreds | Tens | Units |

122

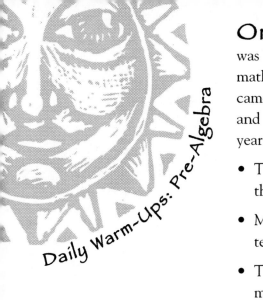

On March 23 of this year, mathematician Emmy Noether was born in Erlangen, Germany. She received a doctoral degree in mathematics in Germany. Forced to leave Germany when the Nazis came into power (she was Jewish), she came to the United States and taught at Bryn Mawr College. Solve this puzzle and learn the year she was born.

- The two-digit number formed by my hundreds and tens digits has these factors: 2, 4, 8, 11, 22, and 44.

- My units digit is the same as the number of triangular faces on a tetrahedron.

- The sum of all of my digits is equal to the sum of the first two multiples of 7.

What year am I?

| _____ | _____ | _____ | _____ |
| Thousands | Hundreds | Tens | Units |

123

On March 31 of this year, Franz Joseph Haydn was born in Rohrau, Austria-Hungary. The "Father of the Symphony," he composed 120 symphonies, more than 100 works for chamber groups, a dozen operas, and hundreds of other musical works. Solve this puzzle to learn the year Haydn was born.

- My hundreds, tens, and units digits are all prime but only my units digit is even.

- My tens digit is 4 less than my hundreds digit.

- The sum of all of my digits is equal to the number of clubs in a pack of playing cards.

What year am I?

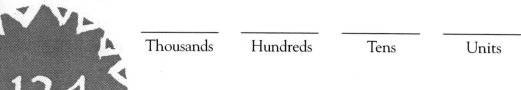

_____   _____   _____   _____

Thousands   Hundreds   Tens   Units

124

© 2003 J. Weston Walch, Publisher

April 1 is the birthdate of Indian political leader, Jagjivan Ram. Born into a family of "untouchables," he was the first of that caste to attend university. A champion and spokesperson for India's 100 million untouchables, he overcame most of the handicaps of the caste system. Learn the year Ram was born by solving this puzzle.

- The two-digit number formed by my hundreds and tens digits is equal to the sum of the even numbers greater than 0 and less than 20.

- My units digit is an even cubic number.

- The sum of all of my digits is equal to the sum of the first three multiples of 3.

What year am I?

_____   _____   _____   _____
Thousands   Hundreds   Tens   Units

125

On April 5 of this year, African-American educator and leader Booker T. Washington was born in Franklin County, Virginia. He wrote, "No race can prosper till it learns that there is as much dignity in tilling a field as in writing a poem." Solve this puzzle to learn the year he was born.

- The two-digit number formed by my tens and units digits has these prime factors: $2^3 \times 7$.

- My hundreds digit is 2 greater than my units digit.

- The sum of all of my digits is equal to the sum of the first four multiples of 2.

    What year am I?

126

_____   _____   _____   _____
Thousands   Hundreds   Tens   Units

On April 10 of this year, Frances Perkins was born in Boston, Massachusetts. The first woman member of a U.S. presidential cabinet, she was appointed Secretary of Labor by Franklin Delano Roosevelt. Learn the year she was born by solving this puzzle.

- The two-digit number formed by my hundreds and tens digits is equal to $8 + 3 \times 40 \div 2 + 10 \times 2$.
- My units digit is equal to $\sqrt[3]{8}$.
- The sum of all of my digits is equal to the eighth prime number.

What year am I?

_____ _____ _____ _____
Thousands   Hundreds   Tens   Units

127

On April 14 of this year, Anne Sullivan was born. She was the teacher and companion of Helen Keller, the remarkable woman who was both blind and deaf. To learn the year Sullivan was born, solve this puzzle.

- The two-digit number formed by my tens and units digits is the sum of the integers between 0 and 12.

- My hundreds digit is the same as the number of sides on an octagon.

- The sum of all of my digits is equal to the missing number in this sequence: 1, 3, 6, 10, 15, ___.

   What year am I?

128

_____   _____   _____   _____
Thousands   Hundreds   Tens   Units

On April 15 of this year, mathematician Leonhard Euler was born. He was the first to use the symbol pi (π), and he published a list of 30 pairs of amicable numbers. (Can you find out what makes a number *amicable?*) Solve this puzzle to find the year Euler was born.

- My hundreds, tens, and units digits form a palindrome with a sum of 14.

- My tens digit is the additive identity.

- The sum of all of my digits is equal to the product of the second and third prime numbers.

What year am I?

_____   _____   _____   _____
Thousands   Hundreds   Tens   Units

129

© 2003 J. Weston Walch, Publisher

On April 23 of this year, William Shakespeare was born in Stratford-upon-Avon in England. He wrote at least 36 plays and 154 sonnets. Shakespeare also died on the 23rd of April. Solve this puzzle to learn the year he was born.

- My hundreds, tens, and units digits are consecutive integers but in the wrong order; their mean is 5.

- My tens digit is 1 greater than my hundreds digit; my units digit is 1 less than my hundreds digit.

- The sum of all of my digits is equal to the number of ounces in 1 pound.

What year am I?

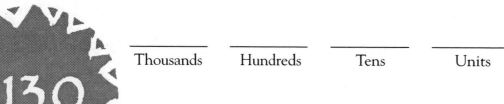

_____   _____   _____   _____
Thousands   Hundreds   Tens   Units

130

On April 26 of this year, Charles Francis Richter was born near Hamilton, Ohio. Richter developed the earthquake magnitude scale that is named after him. If an earthquake has a magnitude of 1 on the Richter scale, it has a power of 10^1; a magnitude of 5 is equal to 10^5 and is 10,000 times more powerful. Find the year he was born.

- The three-digit number formed by my hundreds, tens, and units digits is equal to $2^2 \times 3^2 \times 5^2$.

- The sum of my digits is the same as the number of sides on a decagon.

What year am I?

_____ _____ _____ _____
Thousands   Hundreds   Tens   Units

131

On April 30 of this year, German mathematician Carl Friedrich Gauss was born.

When young Gauss was ordered by his teacher to find the sum of all of the numbers from 1 to 100, he found the answer in a flash by recognizing this pattern:

$$1 + 2 + 3 + \ldots\ldots\ldots + 98 + 99 + 100$$

$$1 + 100 = 101; \; 2 + 99 = 101; \; 3 + 98 = 101; \ldots$$

Since there are 50 pairs of 101, the sum is 101×50.

- My hundreds, tens, and units digits could be the sides of a cube with a volume of 343 cubic units.

- The sum of all of my digits is equal to the sum of the integers greater than or equal to 4 and less than or equal to 7.

What year am I?

132

_____   _____   _____   _____
Thousands   Hundreds   Tens   Units

On May 12 of this year, English author Edward Lear was born. He is remembered for his limericks, published in a book entitled *Book of Nonsense*. Here's an example:

> There was a young poet named Lear
> Who said it is just as I fear
> Five lines are enough
> For this kind of stuff
> Make a limerick each day of the year.

- The two-digit number formed by my hundreds and tens digits is equal to the sum of the first nine odd integers.

- The sum of all of my integers is one dozen.

What year am I?

_____   _____   _____   _____
Thousands   Hundreds   Tens   Units

133

On May 12 of this year, English nurse Florence Nightingale was born. She is considered the person most responsible for making nursing a respected profession. Learn the year she was born by solving this puzzle.

- The two-digit number formed by my tens and units digits is equal to the sum of the first four multiples of 2.

- My hundreds digit is 4 times the size of my tens digit.

- The sum of all of my digits is the fifth prime number.

What year am I?

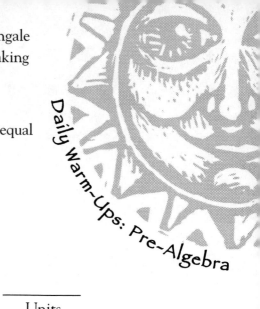

_____   _____   _____   _____
Thousands   Hundreds   Tens   Units

134

On May 15 of this year, author Lyman Frank Baum was born in Chittenango, New York. An American newspaperman, Baum wrote the Wizard of Oz stories. *The Wonderful World of Oz* was the most famous, but he wrote many other books for children, including more than a dozen about Oz. Find the year he was born.

- My units digit is 75% of my hundreds digit.

- My tens and units digits are consecutive integers whose sum is 11 and whose product is 30.

- The sum of all of my digits is equal to the number of pints in 10 quarts.

What year am I?

_____ _____ _____ _____
Thousands   Hundreds   Tens   Units

135

On May 16 of this year, mathematician Maria Gaetana Agnesi was born in Italy. Maria was a gifted child who spoke more than a dozen languages by the age of nine. Her love, however, was mathematics, and even in a time when women were not educated, she became famous for her published books of mathematics. Learn the year she was born by solving this puzzle.

- The two-digit number formed by my tens and units digits (an even number) is 1 greater than the two-digit number formed by my thousands and hundreds digits (a prime number).

- The sum of all of my digits is the seventh prime number.

What year am I?

_____   _____   _____   _____
Thousands   Hundreds   Tens   Units

136

On May 17 of this year, English physician Edward Jenner was born. He is the originator of the inoculation that we use against smallpox. During the Middle Ages, smallpox epidemics caused the deaths of millions of people. Learn the year Jenner was born by solving this puzzle.

- My hundreds digit raised to the 2nd power is equal to the two-digit number formed by my tens and units digits.

- The two-digit number formed by my thousands and hundreds digits is the seventh prime number.

- The sum of all of my digits is equal to the total number of spots on one die.

What year am I?

_____	_____	_____	_____
Thousands	Hundreds	Tens	Units

137

© 2003 J. Weston Walch, Publisher

On May 19 of this year, African-American civil rights leader Malcolm X (Malcolm Little) was born in Omaha, Nebraska. He took the letter "X" to protest the family name assigned by white slave owners to their slaves. He began the Organization of American Unity. He was assassinated in New York City in 1965. Find the year he was born.

- The two-digit number formed by my tens and units digits is equal to $\sqrt{125}$.

- The two-digit number formed by my thousands and hundreds digits is equal to the eighth prime number.

- The sum of all of my digits is equal to 8 more than my hundreds digit.

What year am I?

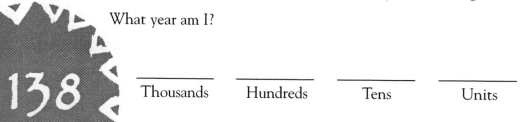

138

_____   _____   _____   _____
Thousands   Hundreds   Tens   Units

On May 23 of this year, German physician Friedrich Anton Mesmer was born. He developed a form of hypnotism, which was called "mesmerism," to treat his patients. Solve this problem to learn the year he was born.

- My tens and units digits are consecutive integers whose sum is 7 and whose product is 12.

- Both my hundreds and tens digits are prime numbers.

- The sum of all of my digits is equal to the sum of the first five counting numbers.

What year am I?

_____ _____ _____ _____
Thousands   Hundreds   Tens   Units

139

On June 5 of this year, the Greek philosopher Socrates was born. His student Plato wrote down Socrates' ideas and teachings. (You need to write B.C.E. after his birth date.)

- My hundreds digit is an even square number; my units digit is an odd square number.

- My tens digit is 150% of my hundreds digit; my units digit is 150% of my tens digit.

- The sum of all of my digits is the eighth prime number.

What year am I?

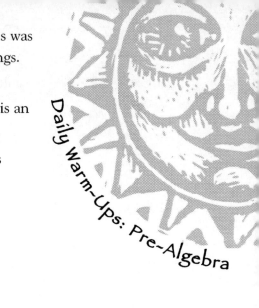

_____   _____   _____

Hundreds   Tens   Units

140

On June 12 of this year, Anne Frank was born. She was the young Jewish girl who kept a diary while hiding in an attic to avoid being captured by the Nazis during World War II. It was later published as a book called *Anne Frank: The Diary of a Young Girl*. Solve this puzzle to learn the year Anne was born.

- My hundreds, tens, and units digits form a palindrome with a sum of 20.

- My tens digit is the only even prime number.

- The sum of all of my digits is equal to the product of the second and fourth prime numbers.

What year am I?

_____   _____   _____   _____
Thousands   Hundreds   Tens   Units

141

On July 22 of this year, American poet Emma Lazarus was born. Words from her poem "The New Colossus" appear at the base of the Statue of Liberty. Some of the words are "Give me your tired, your poor, your huddled masses yearning to breathe free." Find the year she was born by solving this puzzle.

- The two-digit number formed by my tens and units digits is equal to the sum of the first seven odd numbers.

- The sum of my thousands and hundreds digits is equal to my units digit.

- The sum of all of my digits is equal to $6 \times 5 - 4 \times 2$.

  What year am I?

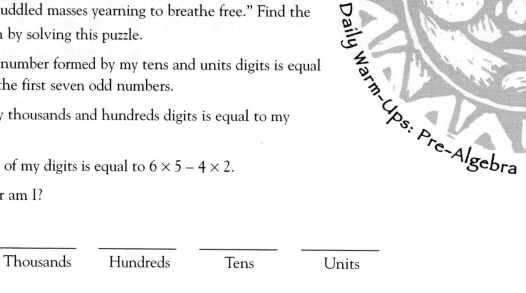

| _____ | _____ | _____ | _____ |
| Thousands | Hundreds | Tens | Units |

142

On September 24 of this year, puppeteer Jim Henson was born in Greenville, Mississippi. Among his creations are Muppets Kermit the Frog, Big Bird, Bert and Ernie, Miss Piggy, the Cookie Monster, and Oscar the Grouch. Find the year Jim Henson was born by solving this puzzle.

- The two-digit number formed by my tens and units digits is equal to $1^3 + 2^3 + 3^3$.

- My hundreds digit is 300% of my tens digit.

- The sum of all of my digits is equal to $4 + 8 \times 3 \div 2 + \sqrt{9}$.

What year am I?

_____	_____	_____	_____
Thousands	Hundreds	Tens	Units

143

© 2003 J. Weston Walch, Publisher

On October 3 of this year, rock-and-roll singer Chubby Checker was born in Philadelphia, Pennsylvania. His real name is Ernest Evans. His most famous song is "The Twist." Solve this puzzle to learn the year he was born.

- The two-digit number formed by my hundreds and tens digits is 6 less than 10^2.

- My units digit is equal to 21^0.

- The sum of all of my digits is the missing number in this sequence: 1, 3, 6, 10, ___, 21, 28.

What year am I?

144

_____   _____   _____   _____
Thousands   Hundreds   Tens   Units

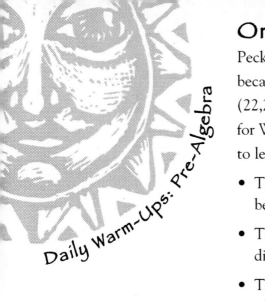

On October 19 of this year, mountain climber Annie Peck was born. She climbed the Matterhorn in the Swiss Alps and became the first American to climb the Peruvian peak Huascarán (22,205 feet). At the age of 61, she placed a banner saying "Votes for Women" at the top of Mount Coropuna in Peru. Solve the puzzle to learn her birth year.

- The two-digit number formed by my tens and units digits would be written L as a Roman numeral.

- The two-digit number formed by my thousands and hundreds digits is equal to $26 + 14 \times 3 - 10 \times 5$.

- The sum of all of my digits is equal to the number of inches in $1\frac{1}{6}$ feet.

What year am I?

| ———— | ———— | ———— | ———— |
| Thousands | Hundreds | Tens | Units |

145

On October 29 of this year, political cartoonist Bill Mauldin was born. Perhaps his most famous cartoon was printed after the assassination of President John F. Kennedy; it portrayed the statue of Lincoln in the Lincoln Memorial in tears. Solve this puzzle to learn the year he was born.

- My tens digit is one greater than my thousands digit.

- The two-digit number formed by my tens and units digits is a multiple of 7.

- The sum of all of my digits is equal to the number of days in $1\frac{6}{7}$ weeks.

What year am I?

_____   _____   _____   _____
Thousands   Hundreds   Tens   Units

146

On November 8 of this year, English astronomer and mathematician Edmund Halley was born. He was the first to observe the comet that is named after him. Halley's Comet can be seen about every 76 years; it is expected to be visible again in 2061. There have been only 28 recorded appearances since 240 B.C.E. Find the year Halley was born.

- The number formed by hundreds and tens digits is equal to the sum of the integers $2 + 3 + 4 + \ldots + 9 + 10 + 11$.

- My units digit is a perfect number.

- The sum of all of my digits is equal to the number of ounces in 1.125 pounds.

What year am I?

Thousands	Hundreds	Tens	Units
_____	_____	_____	_____

147

Historical Highlights

From the first election for president of the United States to the first radio broadcast of *The Lone Ranger*, these puzzles encourage students to explore history from a *mathematical perspective*. Exponents, square roots, number patterns, and more are used to help students find, among other things, the year of the highest wind speed or the last time a new episode of M*A*S*H was seen on TV.

Students will again be using context clues to help them solve these puzzles. Relating historical events to abstract mathematics clues takes the *mystery* out of the problem.

On January 7 of this year, the first presidential election was held in the United States. To learn the year of this historical highlight, solve this puzzle.

- My hundreds, tens, and units digits are consecutive integers whose mean is 8.

- The sum of all of my digits is equal to $3^2 + 4^2$.

What year am I?

——————— ——————— ——————— ———————
Thousands   Hundreds   Tens   Units

148

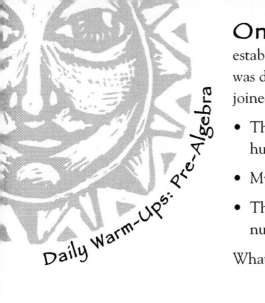

On January 10 of this year, the League of Nations was established. This was the predecessor of the United Nations and was dissolved 26 years after it was formed. The United States never joined. Learn the year it was established by solving this puzzle.

- The sum of my thousands and tens digits is equal to $33\frac{1}{3}\%$ of my hundreds digit.

- My units digit is the additive identity element.

- The sum of all of my digits is one less than the sixth prime number.

What year am I?

_____   _____   _____   _____
Thousands   Hundreds   Tens   Units

149

On January 11 of this year, Chicago schools were closed in the wake of the record-breaking −26°F temperatures. Learn the year by solving this puzzle.

- My units digit raised to the third power is equal to my tens digit.

- My tens digit is 1 less than my hundreds digit.

- The sum of all of my digits is equal to the sum of the first four even numbers starting with 2.

What year am I?

_____   _____   _____   _____
Thousands   Hundreds   Tens   Units

On January 19 of this year, the world's record snowfall fell on London, England. Drifts of 4.5 meters were measured. (How many feet is 4.5 meters?) Solve this puzzle to learn the year.

• My date is a palindrome with a sum of 18.

What year am I?

_____ _____ _____ _____

Thousands   Hundreds   Tens   Units

151

On January 28 of this year, the world watched in horror as the space shuttle *Challenger* exploded 74 seconds into its flight. The seven people who died were teacher Christa McAuliffe and six crew members—Francis Scobee, Michael Smith, Judith Resnick, Ellison Onizuka, Ronald McNair, and Gregory Jarvis. Learn the year of this tragedy by solving this puzzle.

- My units digit is $66\frac{2}{3}\%$ of my hundreds digit but 75% of my tens digit.

- The sum of all of my digits is equal to the first three multiples of 4.

What year am I?

152

_____   _____   _____   _____
Thousands   Hundreds   Tens   Units

On January 30 of this year, the first radio broadcast of *The Lone Ranger* was heard in the United States. The song played at the beginning of the program was the "William Tell Overture" from Rossini's opera. The program started and ended with the phrase "Hi-yo Silver!" Learn the year by solving this puzzle.

- The two-digit number formed by my tens and units digits is equal to $1! + 2! + 3! + 4!$.

- My hundreds digit is equal to the sum of the first three odd integers.

- The sum of all of my digits is equal to 2^4.

What year am I?

_____   _____   _____   _____

Thousands   Hundreds   Tens   Units

153

On February 1 of this year, the United States Supreme Court held its first session in the Royal Exchange Building in New York City. Solve this puzzle to learn the year.

- My hundreds and tens digits are consecutive odd integers with a sum of 16.

- My date is divisible by 5 and 10.

- The sum of all of my digits is the seventh prime number.

What year am I?

_____   _____   _____   _____
Thousands   Hundreds   Tens   Units

154

On February 11 of this year (B.C.E.), the Japanese nation was founded when Emperor Jimmu ascended to the throne. The national holiday is called National Foundation Day, and ceremonies are held with the emperor, empress, and many other dignitaries attending. What is this historical year?

- My three-digit date is divisible by 2, 3, 5, 6, and 10.

- The two-digit number formed by my hundreds and tens digits is equal to the sum of the first 11 counting numbers.

- The sum of all of my digits is equal to $2^2 \times 3$.

What year am I?

_____   _____   _____   _____
Thousands   Hundreds   Tens   Units

155

On February 23 of this year, members of the Fifth United States Marine division planted an American flag atop Mount Suribachi on Iwo Jima. This act was memorialized in a very famous photograph. Learn the year this occurred by solving this puzzle.

- The sum of my consecutive tens and units digits is equal to my hundreds digit.

- My date is divisible by 5.

- The sum of my digits is equal to $18 \div 2 + 60 \div 4 - \sqrt{25}$.

What year am I?

Thousands	Hundreds	Tens	Units
_____	_____	_____	_____

On February 23 of this year, the siege of the Alamo began in San Antonio, Texas. To learn the year, solve this puzzle.

- The two-digit number formed by my tens and units digits is equal to $1^3 + 2^3 + 3^3$.

- The sum of my thousands and hundreds digits is equal to the sum of my tens and units digits.

- The sum of all of my digits is equal to 2×3^2.

What year am I?

Thousands	Hundreds	Tens	Units
_____	_____	_____	_____

157

© 2003 J. Weston Walch, Publisher

On February 28 of this year, the final episode of the TV show M*A*S*H aired. This was the last of the show's 255 episodes and was watched by over 75% of the viewing public. Solve this puzzle to learn the year.

- The two-digit number formed by my hundreds and tens digits is equal to the sum of $1^4 + 2^4 + 3^4$.

- My units digit is the square root of my hundreds digit.

- The sum of my digits is equal to the next number in the Fibonacci sequence: 1, 1, 2, 3, 5, 8, 13, ___.

What year am I?

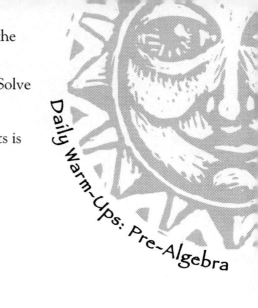

158

_____ _____ _____ _____
Thousands　　Hundreds　　　Tens　　　　Units

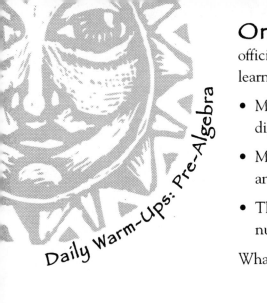

On March 3 of this year, "The Star-Spangled Banner" officially became the national anthem of the United States. To learn the year, solve this puzzle.

- My hundreds and tens digits are both odd numbers—my hundreds digit is 3 times my tens digit.

- My thousands and units digits could be the sides of a square with an area of 1 square unit and a perimeter of 4 units.

- The sum of all of my digits is the sum of the first three square numbers.

What year am I?

_____   _____   _____   _____
Thousands   Hundreds   Tens   Units

On March 10 of this year, American abolitionist Harriet Tubman died in Auburn, New York. A former slave, Tubman escaped from a Maryland plantation and formed the "Underground Railroad." Through her efforts, more than 300 slaves reached freedom. Solve this puzzle to learn the year of her death.

- The two-digit number formed by my hundreds and tens digits is equal to $1^2 + 2^2 + 3^2 + 4^2 + 5^2 + 6^2$.

- My units digit is $33\frac{1}{3}\%$ of my hundreds digit.

- The sum of all of my digits is the product of the first and fourth prime numbers.

What year am I?

_____	_____	_____	_____
Thousands	Hundreds	Tens	Units

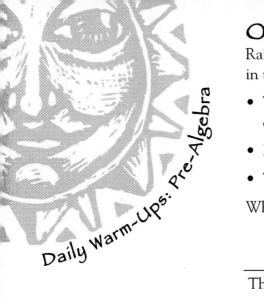

On March 10 of this year, Commissioner George Scott Railaton and seven women officers established the Salvation Army in the United States. To learn the year, solve this puzzle.

- The two-digit number formed by my hundreds and tens digits is equal to $2^3 \times 11$.

- My date is divisible by 2, 5, and 10.

- The sum of all of my digits is equal to $2^4 + 1^4$.

What year am I?

_____   _____   _____   _____
Thousands   Hundreds   Tens   Units

161

© 2003 J. Weston Walch, Publisher

On March 20 of this year, children rang the United Nations bell in New York City for the first celebration of Earth Day. The bell sounded at the exact moment when the sun crossed the equator, the beginning of spring in the Northern Hemisphere. To learn the year, solve this puzzle.

- My tens and units digits are consecutive odd integers; they could be the sides of a rectangle with a perimeter of 32 and an area of 63.

- My hundreds digit is equal to the sum of the integers between 1 and 5.

- The sum of all of my digits is equal to the number of letters in the English alphabet.

What year am I?

162

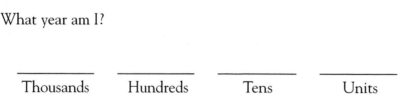

| _____ | _____ | _____ | _____ |
| Thousands | Hundreds | Tens | Units |

On March 23 of this year, Patrick Henry delivered his "Give me liberty or give me death" speech. To learn the year, solve this puzzle.

- The two-digit number formed by my tens and units digits is equal to the sum of the integers greater than 2 and less than 13.
- My hundreds digit is a Mersenne prime; it is equal to $2^3 - 1$.
- The sum of all of my digits is equal to two decades.

What year am I?

_____   _____   _____   _____

Thousands   Hundreds   Tens   Units

On April 11 of this year, the American Society for the Prevention of Cruelty to Animals (A.S.P.C.A.) was chartered in New York State. To learn the year, solve this puzzle.

- My tens and units digits are the same number; each one is $\frac{3}{4}$ of my hundreds digit.

- My hundreds digit is the same as the number of vertices on a cube.

- The sum of all of my digits is equal to $\sqrt{49} \times \sqrt{9}$.

What year am I?

_____   _____   _____   _____
Thousands   Hundreds   Tens   Units

164

On April 12 of this year, the wind speed reached the highest ever recorded in the United States. At the Mount Washington Observatory in New Hampshire, weather observers Wendell Stevenson, Alexander McKenzie, and Salvatore Pagliuca observed and recorded gusts over 231 miles per hour.

- My tens and units digits are consecutive integers whose sum is 7.
- My hundreds digit is 300% of my tens digit.
- The sum of all of my digits is the seventh prime number.

What year am I?

_____	_____	_____	_____
Thousands	Hundreds	Tens	Units

165

On April 17 of this year, the 355-year "state of war" that had existed between the Netherlands and the Scilly Isles came to an end. The Dutch ambassador flew to the Scilly Isles to deliver a proclamation terminating the war. Though hostilities had ended centuries before, no one had bothered to declare the war over. Learn the year of this truce by solving the puzzle.

- My units digit is a perfect number.
- My tens and units digits are consecutive even integers with a product of 48.
- My units digit is $\frac{2}{3}$ of my hundreds digit.
- The sum of all of my digits is equal to 4!.

What year am I?

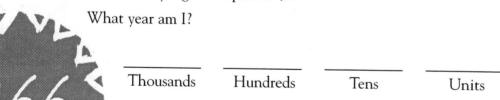

| _____ | _____ | _____ | _____ |
| Thousands | Hundreds | Tens | Units |

166

On May 6 of this year, the dirigible Hindenburg exploded as it approached its mooring in New Jersey. It had just completed a transatlantic crossing. Thirty-six of its ninety-seven passengers and crew died. To learn the year this occurred, solve this puzzle.

• My tens digit is $\frac{1}{3}$ of my hundreds digit.
• The two-digit number formed by my tens and units digits is the largest prime number less than 40.
• The sum of all of my digits is the same as the number of faces on an icosahedron.

What year am I?

| _____ | _____ | _____ | _____ |
| Thousands | Hundreds | Tens | Units |

167

On May 7 of this year, a German submarine torpedoed the British passenger ship *Lusitania* traveling from New York to England. It was carrying nearly 2,000 passengers; 1,198 lives were lost. Germany had warned President Wilson in advance, claiming that the *Lusitania* was carrying ammunition to England. Learn the year of this disaster by solving this puzzle.

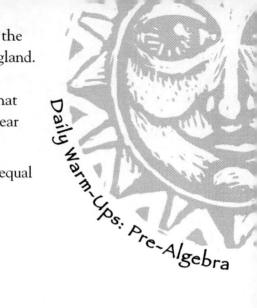

- The two-digit number formed by my tens and units digits is equal to the Mersenne prime $2^4 - 1$.

- My hundreds digit is a square number.

- The sum of all of my digits is equal to 4^2.

What year am I?

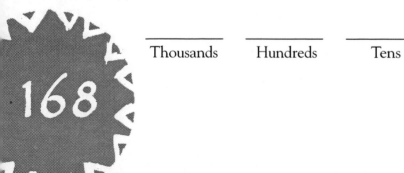

| _____ | _____ | _____ | _____ |
| Thousands | Hundreds | Tens | Units |

On June 2 of this year, Congress granted Native Americans citizenship. Learn the year by solving this puzzle.

- The two-digit number formed by my tens and units digits is a multiple of 2, 3, 4, 6, and 12.

- My hundreds digit is 1 more than twice my units digit.

- The sum of all of my digits is the same as the number of ounces in 1 pound.

What year am I?

_____   _____   _____   _____
Thousands   Hundreds   Tens   Units

169

On June 15 of this year, England's King John signed the Magna Carta. This is considered to be one of the most important documents in the history of humanity's search for liberty and freedom. Four copies have survived through the years. Learn the year of the signing by solving this puzzle.

- The two-digit number formed by my thousands and hundreds digits is the same as the number of sides on a dodecagon.

- The two-digit number formed by my tens and units digits is the missing number in this series: 1, 3, 6, 10, ___, 21, 28 . . .

- The sum of all of my digits is equal to $\sqrt{100} - 1$.

What year am I?

_____   _____   _____   _____
Thousands   Hundreds   Tens   Units

170

Daily Warm-Ups: Pre-Algebra

On July 17 of this year, Douglas Corrigan flew out of Brooklyn, New York, heading for Los Angeles, California. Twenty-eight hours later, he landed in Dublin, Ireland. When he arrived back in the United States, he was nicknamed "Wrong Way" Corrigan. Learn the year of his flight by solving this puzzle.

- The two-digit number formed by my tens and units digits is equal to the product of the first and eighth prime numbers.

- The sum of my thousands and units digits is equal to my hundreds digit.

- The sum of all of my digits is $4^2 + 2^2 + 1^2$.

What year am I?

_____   _____   _____   _____
Thousands   Hundreds   Tens   Units

171

On August 5 of this year, the first English colony in North America was founded by Sir Humphrey Gilbert. It was established in the area around St. John's Harbor, Newfoundland. Find the year by solving this puzzle.

- The two-digit number formed by my hundreds and tens digits is equal to the sum of the integers $13 + 14 + 15 + 16$.

- My units digit is a prime number; it is 2 less than my hundreds digit.

- The sum of all of my digits is $4! - 3! - 1!$.

What year am I?

_____ _____ _____ _____
Thousands   Hundreds   Tens   Units

172

On August 7 of this year, the first picture of Earth was received from the space satellite *Explorer VI*. It was the first time the people of Earth could see their planet from space. Learn the year by solving this problem.

- My hundreds, tens, and units digits form a palindrome with a sum of 23.

- The two-digit number formed by my thousands and hundreds digits is equal to $\sqrt{25} \times \sqrt{16} - \sqrt{1}$.

What year am I?

_____   _____   _____   _____
Thousands   Hundreds   Tens   Units

173

On September 23 of this year, the planet Neptune was first observed. Neptune, the eighth planet from the sun, is about 2.8 billion miles from the sun; it takes 165 years to revolve around the sun and has a diameter of about 30,000 miles. (How do these statistics compare with Earth's?) Find the year of this observation.

- My units digit is 2 more than my tens digit; my tens digit is $\frac{1}{2}$ of my hundreds digit.

- My units digit is equal to 3!.

- The sum of my digits is 1 less than two decades.

What year am I?

| Thousands | Hundreds | Tens | Units |

174

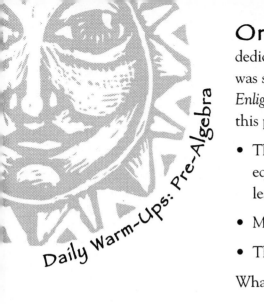

On October 28 of this year, the Statue of Liberty was dedicated on Bedloe's Island in the harbor of New York City. It was sculpted by Frédéric-Auguste Bartholdi who called it *Liberty Enlightening the World.* Learn the year of the dedication by solving this puzzle.

- The two-digit number formed by my tens and hundreds digits is equal to the sum of the integers greater than or equal to 3 and less than or equal to 13.

- My units digit is a perfect number.

- The sum of all of my digits is the ninth prime number.

What year am I?

Thousands	Hundreds	Tens	Units
_____	_____	_____	_____

175

On November 9 of this year, New York City and most of the northeastern United States was blacked out due to an electric power failure. More than 30 million people, in an 80,000-square-mile area, were affected. Solve this puzzle to learn the year that this happened.

- The sum of my thousands and units digits is equal to my perfect tens digit.

- The product of my hundreds digit and units digit is 45.

- The sum of all of my digits is equal to the 8th number in the Fibonacci sequence.

What year am I?

176

| _____ | _____ | _____ | _____ |
| Thousands | Hundreds | Tens | Units |

Daily Warm-Ups: Pre-Algebra

On November 18 of this year, the squeaky-voiced Mickey Mouse first appeared on a movie screen at the Colony Theater in New York City. Walt Disney's *Steamboat Willie* was the first animated cartoon talking picture. Learn the year of this movie breakthrough by solving this puzzle.

- My tens and units digits are Fibonacci numbers whose sum is 10 and product is 16.

- My tens digit is 25% of my units digit.

- The difference between my thousands and hundreds digits is equal to my units digit.

- The sum of my digits is divisible by 2, 4, 5, and 10.

What year am I?

_____ _____ _____ _____
Thousands   Hundreds   Tens   Units

© 2003 J. Weston Walch, Publisher

On November 19 of this year, President Abraham Lincoln delivered the Gettysburg Address at a ceremony dedicating 17 acres of the battlefield in Gettysburg, Pennsylvania. Though the speech took less than two minutes, it is recognized as one of the most eloquent in the English language. Find the year by solving this puzzle.

- My prime units digit is $\frac{1}{2}$ of my perfect tens digit.

- My date is divisible by both 3 and 9.

- The sum of my thousands and hundreds digits is equal to the sum of my tens and units digits.

What year am I?

| _____ | _____ | _____ | _____ |
| Thousands | Hundreds | Tens | Units |

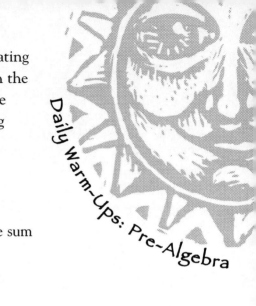

178

On December 15 of this year, the Bill of Rights, the first 10 amendments to our Constitution, became effective following the approval of Virginia. Find the year of this historical highlight by solving this puzzle.

- My hundreds and tens digits are consecutive odd integers whose sum is 16; my hundreds digit is prime.

- My units digit is the multiplicative identity element.

- The sum of all of my digits is the same as the number of inches in 1.5 feet.

What year am I?

_____ _____ _____ _____
Thousands   Hundreds   Tens   Units

179

On December 1 of this year, Rosa Parks was arrested in Montgomery, Alabama, for refusing to give up her seat and move to the back of the bus. Her arrest triggered a yearlong boycott of the city bus system and helped end racial segregation on municipal buses in the South. This event has been called the beginning of the Civil Rights Movement in the United States.

- My tens and units digits are the same number; their sum is equal to the sum of my thousands and hundreds digits.

- The sum of all of my digits is equal to $8 \times 5 \div \sqrt[3]{8}$.

What year am I?

_____   _____   _____   _____
Thousands   Hundreds   Tens   Units

States of the Union

Page 1: The Constitution was adopted in 1787.
Page 2: Delaware, Pennsylvania, and New Jersey became states in 1787.
Page 3: Georgia, Massachusetts, Connecticut, New Hampshire, South Carolina, Virginia, New York, and Maryland all became states in 1788.
Page 4: North Carolina became a state in 1789.
Page 5: Rhode Island became a state in 1790.
Page 6: Vermont became a state in 1791.
Page 7: Kentucky became a state in 1792.
Page 8: Tennessee became a state in 1796.
Page 9: Ohio became a state in 1803.
Page 10: Louisiana became a state in 1812.
Page 11: Indiana became a state in 1816.
Page 12: Mississippi became a state in 1817.
Page 13: Illinois became a state in 1818.
Page 14: Alabama became a state in 1819.
Page 15: Maine became a state in 1820.
Page 16: Missouri became a state in 1821.
Page 17: Arkansas became a state in 1836.
Page 18: Michigan became a state in 1837.
Page 19: Florida and Texas became states in 1845.
Page 20: Iowa became a state in 1846.
Page 21: Wisconsin became a state in 1848.
Page 22: California became a state in 1850.
Page 23: Minnesota became a state in 1858.
Page 24: Oregon became a state in 1859.
Page 25: Kansas became a state in 1861.
Page 26: West Virginia became a state in 1863.
Page 27: Nevada became a state in 1864.
Page 28: Nebraska became a state in 1867.
Page 29: Colorado became a state in 1876.
Page 30: North Dakota, South Dakota, Montana, and Washington became states in 1889.
Page 31: Idaho and Wyoming became states in 1890.
Page 32: Utah became a state in 1896.
Page 33: Oklahoma became a state in 1907.
Page 34: New Mexico and Arizona became states in 1912.
Page 35: Alaska and Hawaii became states in 1959.

Famous Firsts

Page 36: The women became FBI agents in 1972.
Page 37: The first junior high school opened in 1910.

Daily Warm-Ups: Pre-Algebra

Page 38:	Nellie Tayloe Ross became the first woman governor in 1925.
Page 39:	Blanchard flew the balloon in 1793.
Page 40:	Hattie Caraway became the first woman U.S. Senator in 1932.
Page 41:	Robert C. Weaver became a member of Johnson's cabinet in 1966.
Page 42:	Driver's licenses were required in 1937.
Page 43:	Dr. Blackwell received her degree in 1849.
Page 44:	Justice Brandeis was appointed in 1916.
Page 45:	*These Are My Children* aired in 1949.
Page 46:	Maggie Walker became a bank manager in 1903.
Page 47:	The first singing telegram was delivered in 1933.
Page 48:	*The New York Times* slogan was published in 1897.
Page 49:	Glenn orbited the earth in 1962.
Page 50:	The school for the blind was opened in 1829.
Page 51:	The first auto was driven in 1896.
Page 52:	The parachute jump was made in 1912.
Page 53:	*Freedom's Journal* was first published in 1827.
Page 54:	The swallows first returned in 1776.
Page 55:	The first women's college basketball game was played in 1893.
Page 56:	Hank Aaron hit the home run in 1974.
Page 57:	Robinson was recruited in 1947.
Page 58:	The Battle of Lexington took place in 1775.
Page 59:	The first Seeing Eye dog was presented in 1923.
Page 60:	Gwendolyn Brooks won the Pulitzer Prize in 1950.
Page 61:	De Soto reached the Mississippi River in 1541.
Page 62:	The railroad was completed in 1869.
Page 63:	The Alder Planetarium was opened in 1930.
Page 64:	Sally Ride's mission was in 1983.
Page 65:	Neil Armstrong landed on the moon in 1969.
Page 66:	The balloon crossing occurred in 1978.
Page 67:	The baby was born in 1893.
Page 68:	The newspaper was published in 1690.
Page 69:	Thurgood Marshall became a justice in 1967.

Discoveries, Inventions, and Notable Accomplishments

Page 70:   The X ray was discovered in 1895.

Page 71:   "The Landlord's Game" was patented in 1904.

Page 72:   The accordion was patented in 1854.

Page 73:   The Pentagon was completed in 1943.

Page 74:   Gold was discovered in 1848.

Page 75:   The phone call was made in 1877.

Page 76:   The tomb was opened in 1923.

Page 77:   The phonograph was invented in 1878.

Page 78:   The electric razor was patented in 1931.

Page 79:   Paper money became legal tender in 1862.

Page 80:   Dr. Jonas Salk introduced the polio vaccine in 1953.

Page 81:   Teflon® was invented in 1938.

Page 82:   Radium was isolated in 1902.

Page 83:   The Barbie® doll was invented in 1959.

Page 84:   The disposable diaper was invented in 1950.

Page 85:   Play-Doh® was invented in 1956.

Page 86:   Sarah Boone invented her ironing board in 1892.

Page 87:   The Slinky® was invented in 1945.

Page 88:   The Simplon Tunnel was opened in 1906.

Page 89:   Lindbergh crossed the Atlantic in 1927.

Page 90:   The Brooklyn Bridge was opened in 1883.

Page 91:   Liquid paper was invented in 1956.

Page 92:   The Dionne quintuplets were born in 1934.

Page 93:   "Casey at the Bat" was printed in 1888.

Page 94:   The vacuum cleaner was patented in 1869.

Page 95:   Benjamin Franklin conducted his experiment in 1752.

Page 96:   The wheat reaper was patented in 1834.

Page 97:   Patsy Sherman discovered Scotchgard Fabric Protector in 1952.

Page 98:   The bar code was patented in 1952.

Page 99:   The Holland Tunnel was opened in New York City in 1927.

Page 100: The *Spruce Goose* made its flight in 1947.

Page 101: Evaporated milk was patented in 1884.

Page 102: The painting was hung in 1961.

Page 103: The Louisiana Purchase took place in 1803.

Happy Birthday to You ...

Page 104: Paul Revere was born in 1735.

Page 105:   J.R.R. Tolkien was born in 1892.

Answer Key

Page 106:	Dr. Martin Luther King, Jr., was born in 1929.
Page 107:	A.A. Milne was born in 1882.
Page 108:	Maria Tallchief was born in 1925.
Page 109:	Mozart was born in 1756.
Page 110:	Hank Aaron was born in 1934.
Page 111:	Laura Ingalls Wilder was born in 1867.
Page 112:	Grant Woods was born in 1892.
Page 113:	Galileo was born in 1564.
Page 114:	Susan B. Anthony was born in 1820.
Page 115:	Shalom Aleichem was born in 1859.
Page 116:	Copernicus was born in 1473.
Page 117:	Renoir was born in 1841.
Page 118:	Dr. Seuss was born in 1904.
Page 119:	Michelangelo was born in 1475.
Page 120:	Luther Burbank was born in 1849.
Page 121:	Einstein was born in 1879.
Page 122:	Pablo Juárez was born in 1806.
Page 123:	Emmy Noether was born in 1884.
Page 124:	Haydn was born in 1732.
Page 125:	Jagjivan Ram was born in 1908.
Page 126:	Booker T. Washington was born in 1856.
Page 127:	Frances Perkins was born in 1882.
Page 128:	Anne Sullivan was born in 1866.
Page 129:	Leonhard Euler was born in 1707.
Page 130:	Shakespeare was born in 1564.
Page 131:	Charles Richter was born in 1900.
Page 132:	Carl Gauss was born in 1777.
Page 133:	Edward Lear was born in 1812.
Page 134:	Nightingale was born in 1820.
Page 135:	Frank Baum was born in 1856.
Page 136:	Maria Agnesi was born in 1718.
Page 137:	Edward Jenner was born in 1749.
Page 138:	Malcolm X was born in 1925.
Page 139:	Friedrich Mesmer was born in 1734.
Page 140:	Socrates was born in 469 B.C.E.
Page 141:	Anne Frank was born in 1929.
Page 142:	Emma Lazarus was born in 1849.
Page 143:	Jim Henson was born in 1936.
Page 144:	Chubby Checker was born in 1941.
Page 145:	Annie Peck was born in 1850.
Page 146:	Bill Mauldin was born in 1921.
Page 147:	Edmund Halley was born in 1656.

Historical Highlights

Page 148: The first election was held in 1789.

Page 149: The League of Nations was established in 1920.

Page 150: The schools closed in 1982.

Page 151: The record snowfall occurred in 1881.

Page 152: *Challenger* exploded in 1986.

Page 153: *The Lone Ranger* was heard in 1933.

Page 154: The Supreme Court first met in 1790.

Page 155: Japan was founded in 660 B.C.E.

Page 156: The Marines planted the flag in 1945.

Page 157: The siege of the Alamo began in 1836.

Page 158: The last episode of M*A*S*H aired in 1983.

Page 159: "The Star-Spangled Banner" became our national anthem in 1931.

Page 160: Harriet Tubman died in 1913.

Page 161: The Salvation Army was established in 1880.

Page 162: Earth Day was founded in 1979.

Page 163: Patrick Henry gave his speech in 1775.

Page 164: The A.S.P.C.A. was founded in 1866.

Page 165: The record wind occurred in 1934.

Page 166: The 335-year war ended in 1986.

Page 167: The *Hindenburg* exploded in 1937.

Page 168: The *Lusitania* was sunk in 1915.

Page 169: Native Americans were granted citizenship in 1924.

Page 170: The Magna Carta was signed in 1215.

Page 171: "Wrong Way" Corrigan made the flight in 1938.

Page 172: The English colony was founded in 1583.

Page 173: Pictures of Earth were seen in 1959.

Page 174: The planet Neptune was observed in 1846.

Page 175: The Statue of Liberty was given to the United States in 1886.

Page 176: The blackout occurred in 1965.

Page 177: Mickey Mouse debuted in 1928.

Page 178: The Gettysburg Address was delivered in 1863.

Page 179: Rosa Parks took her stand in 1955.

Page 180: The Bill of Rights became effective in 1791.

Daily Warm-Ups: Pre-Algebra

additive identity element—a number that when added to any given number always results in a sum identical to the given number. In the real number system, the additive identity is 0.

area—the number of square units needed to cover a surface. The formulas for finding some common areas are

Rectangle:	$A = \text{length} \times \text{width}$
Triangle:	$A = \dfrac{\text{base} \times \text{height}}{2}$
Trapezoid:	$A = \dfrac{1}{2}[\text{height}(\text{base}_1 + \text{base}_2)]$

average (mean)—the measure of central tendency found by finding the sum of the data and dividing by the number of terms

consecutive integers—a series of two or more integers in which each is one greater than the preceding integer, e.g., 7, 8, 9 are consecutive integers

decagon—a two-dimensional polygon with 10 sides and 10 angles

dodecahedron—a three-dimensional polyhedron comprised of 12 pentagons. It is one of the five Platonic solids with three pentagons meeting at each vertex.

factor—When two or more numbers are multiplied, each number is a factor of the product. The factors of 48 are 1, 2, 3, 4, 6, 8, 12, 16, 24, and 48.

factorial (!)—the product of a given series of consecutive whole numbers beginning with 1 and ending with the number

$$0! = 1$$
$$1! = 1 \times 1 = 1$$
$$2! = 2 \times 1 = 2$$
$$3! = 3 \times 2 \times 1 = 6$$
$$n! = n \times (n-1) \times (n-2) \times (n-3) \ldots \times 1$$

Fibonacci sequence—an infinite sequence of natural numbers in which each term is the sum of the preceding two terms: 1, 1, 2, 3, 5, 8, 13, 21 . . .

heptagon—a two-dimensional polygon with 7 sides and 7 angles

icosahedron—a three-dimensional polyhedron comprised of 20 equilateral triangles. It is one of the five Platonic Solids with five triangles at each vertex.

Mersenne prime—a prime in the form $2^p - 1$ where p is a prime. Thirty-one is a Mersenne prime because $2^5 - 1 = 31$. Seven is a Mersenne prime because $2^3 - 1 = 7$.

multiplicative identity element—a number that when multiplied by any given number always results in a product identical to the given number. In the real number system, the multiplicative identity is 1.

nonagon—a two-dimensional polygon with nine angles and nine sides

octagon—a two-dimensional polygon with eight sides and eight angles

order of operations—the order in which arithmetic operations must be performed by definition. The following rules have been agreed upon:

1. First, do all operations within parentheses.

2. Then do all powers and roots in order from left to right.

3. Then do all multiplication and division in order from left to right.

4. Then do all addition and subtraction in order from left to right.

palindrome—a natural number of two or more digits that is the same whether it is read forwards or backwards, example: 44, 121, 1331, 1234321

perfect number—a natural number that equals the sum of its proper divisors—that is, all of its whole number divisors except the number itself

Examples: 6 is a perfect number because
$$1 + 2 + 3 = 6$$
28 is a perfect number because
$$1 + 2 + 4 + 7 + 14 = 28$$
The next two perfect numbers are 496 and 8,128.

prime factorization—the process of finding all the prime factors of a number. The *fundamental theorem of arithmetic* states that every integer greater than one can be expressed as a product of prime factors in one and only one way, except for the order of the factors.

Example: The prime factorization of 12 is $2 \times 2 \times 3$.

prime number—any natural number greater than one that has two and only two factors. The prime numbers less than 100 are 2, 3, 5, 7, 11, 13, 17, 19, 23, 29, 31, 37, 41, 43, 47, 53, 59, 61, 67, 71, 73, 79, 83, 89, 97.

rules of divisibility—a number is divisible by

2: if the units digit is 0, 2, 4, 6, or 8; if it is an even number

3: if the sum of the digits of the number is divisible by 3

4: if the last two digits of the number are divisible by 4

5: if the last digit of the number is 0 or 5

6: if the number is even and is divisible by 3

8: if the last three digits of the number are divisible by 8

9: if the sum of the digits of the number is divisible by 9

10: if the last digit of the number is a 0

square number—the product of a number multiplied by itself. One, 4, 16, 25 . . . n^2 are all square numbers.

tetrahedron—a three-dimensional polyhedron comprised of four equilateral triangles. One of the five Platonic Solids, a tetrahedron has three triangles at each vertex.

volume—the number of cubic units necessary to fill a three-dimensional space. The volume of a cube = side^3.

Turn downtime into learning time!

Other books in the Daily *Warm-Ups* series:

- Algebra
- Analogies
- Biology
- Critical Thinking
- Earth Science
- Geography
- Geometry
- Journal Writing
- Poetry
- Shakespeare
- Spelling & Grammar
- Test-Prep Words
- U.S. History
- Vocabulary
- Writing
- World History